BURLEIGH DODDS SCIENCE: INSTANT INSIGHTS

NUMBER 110

Alternative sources of protein for pigs

I0130564

burleigh dodds
SCIENCE PUBLISHING

Published by Burleigh Dodds Science Publishing Limited
82 High Street, Sawston, Cambridge CB22 3HJ, UK
www.bdspublishing.com

Burleigh Dodds Science Publishing, 1518 Walnut Street, Suite 900, Philadelphia, PA 19102-3406, USA

First published 2025 by Burleigh Dodds Science Publishing Limited
© Burleigh Dodds Science Publishing, 2025. All rights reserved.

British Library Cataloguing in Publication Data
A catalogue record for this book is available from the British Library

ISBN 978-1-83545-012-3 (Print)
ISBN 978-1-83545-013-0 (ePub)

DOI: 10.19103/9781835450130

Typeset by Deanta Global Publishing Services, Dublin, Ireland

Contents

Acknowledgements ix

1 Developing seaweed/macroalgae as feed for pigs 1
Marta López-Alonso, Universidade de Santiago de Compostela,
Spain; Marco García-Vaquero, University College Dublin, Ireland;
and Marta Miranda, Universidade de Santiago de Compostela,
Spain

  1 Introduction 1
  2 Challenges in using macroalgae for feed applications 2
  3 Composition of macroalgae 4
  4 Biological functions and health-promoting effects of macroalgae and
    macroalgal-derived extracts in pig nutrition 10
  5 Conclusion and future trends 16
  6 Where to look for further information 17
  7 Acknowledgements 18
  8 References 18

2 High protein corn fermentation products for swine derived
  from corn ethanol production 27
Peter E. V. Williams, FluidQuipTechnologies, USA

  1 Introduction 27
  2 Distillers dried grains with solubles 28
  3 Corn fermented protein 29
  4 Challenges in producing corn fermented protein 31
  5 Case study: standardized ileal digestibility of CFP for pigs 33
  6 Case study: concentrations of digestible and metabolizable energy in
    CFP products fed to pigs 35
  7 Case study: effects of adding phytase on availability of calcium and
    phosphorus in corn-fermented products 40
  8 Case study: effects on performance of pigs of inclusion of corn-
    fermented protein in the diet 43

9   Conclusion   46
10 References   48

3   Developing alternative sources of protein in pig nutrition: insects   51
    Kristy DiGiacomo, The University of Melbourne, Australia

    1   Introduction   51
    2   Why insects   52
    3   Current state of insect production   53
    4   Potential insects and current production volumes   59
    5   Production and health responses in pigs   60
    6   Competition for resources   61
    7   Rearing insects on manure   63
    8   Current research   64
    9   Future trends in research   65
    10 Barriers to uptake/challenges to production   66
    11 Conclusion   67
    12 Where to look for further information   67
    13 References   68

4   Black soldier fly meal: an alternative protein source for pigs   83
    S. Struthers, and J. G. M. Houdijk, Scotland's Rural College (SRUC),
    UK; and H. N. Hall, Anpario plc, UK

    1   Introduction   83
    2   Nutritional composition of black soldier fly   84
    3   Production performance of pigs fed diets containing black soldier fly larvae 86
    4   Benefits of using black soldier fly   93
    5   Challenges   97
    6   Applications   98
    7   Conclusion   99
    8   Acknowledgements   99
    9   References   100

Series list

Title	Series number
Sweetpotato	01
Fusarium in cereals	02
Vertical farming in horticulture	03
Nutraceuticals in fruit and vegetables	04
Climate change, insect pests and invasive species	05
Metabolic disorders in dairy cattle	06
Mastitis in dairy cattle	07
Heat stress in dairy cattle	08
African swine fever	09
Pesticide residues in agriculture	10
Fruit losses and waste	11
Improving crop nutrient use efficiency	12
Antibiotics in poultry production	13
Bone health in poultry	14
Feather-pecking in poultry	15
Environmental impact of livestock production	16
Sensor technologies in livestock monitoring	17
Improving piglet welfare	18
Crop biofortification	19
Crop rotations	20
Cover crops	21
Plant growth-promoting rhizobacteria	22
Arbuscular mycorrhizal fungi	23
Nematode pests in agriculture	24
Drought-resistant crops	25
Advances in detecting and forecasting crop pests and diseases	26
Mycotoxin detection and control	27
Mite pests in agriculture	28
Supporting cereal production in sub-Saharan Africa	29
Lameness in dairy cattle	30
Infertility and other reproductive disorders in dairy cattle	31
Alternatives to antibiotics in pig production	32
Integrated crop–livestock systems	33
Genetic modification of crops	34

Developing forestry products   35

Reducing antibiotic use in dairy production   36

Improving crop weed management   37

Improving crop disease management   38

Crops as livestock feed   39

Decision support systems in agriculture   40

Fertiliser use in agriculture   41

Life cycle assessment (LCA) of crops   42

Pre- and probiotics in poultry production   43

Poultry housing systems   44

Ensuring animal welfare during transport and slaughter   45

Conservation tillage in agriculture   46

Tropical forestry   47

Soil health indicators   48

Improving water management in crop cultivation   49

Fungal diseases of apples   50

Using crops as biofuel   51

Septoria tritici blotch in cereals   52

Biodiversity management practices   53

Soil erosion   54

Integrated weed management in cereal cultivation   55

Sustainable forest management   56

Restoring degraded forests   57

Developing immunity in pigs   58

Bacterial diseases affecting pigs   59

Viral diseases affecting pigs   60

Developing immunity in poultry   61

Managing bacterial diseases of poultry   62

Proximal sensors in agriculture   63

Dietary supplements in dairy cattle nutrition   64

Dietary supplements in poultry nutrition   65

Intercropping   66

Managing arthropod pests in tree fruit   67

Weed management in regenerative agriculture   68

Integrated pest management in cereal cultivation   69

Economics of key agricultural practices   70

Nutritional benefits of milk   71

Biostimulant applications in agriculture 72

Phosphorus uptake and use in crops 73

Optimising pig nutrition 74

Nutritional and health benefits of beverage crops 75

Artificial Intelligence applications in agriculture 76

Ensuring the welfare of laying hens 77

Ecosystem services delivered by forests 78

Improving biosecurity in livestock production 79

Rust diseases of cereals 80

Optimising rootstock health 81

Irrigation management in horticultural production 82

Improving the welfare of gilts and sows 83

Improving the shelf life of horticultural produce 84

Improving the health and welfare of heifers and calves 85

Managing arthropod pests in cereals 86

Optimising photosynthesis in crops 87

Optimising quality attributes in poultry products 88

Advances in fertilisers and fertiliser technology 89

Sustainable tropical forest management 90

Phenotyping applications in agriculture 91

Fungicide resistance in cereals 92

Unmanned aircraft systems in agriculture 93

Using manure in soil management 94

Zero/no till cultivation 95

Optimising agri-food supply chains 96

Optimising quality attributes in horticultural products 97

Improving the sustainability of dairy production 98

Improving the welfare of growing and finishing pigs 99

Ensuring the welfare of broilers 100

Biofertiliser use in agriculture 101

Optimising reproductive efficiency in pigs 102

Regenerative techniques to improve soil health 103

Agroforestry practices 104

Carbon monitoring and management in forests 105

Viruses affecting horticultural crops 106

Regulatory frameworks for new agricultural products and technologies 107

Machine vision applications in agriculture 108

Alternative sources of protein for poultry   109

Alternative sources of protein for pigs   110

Nitrogen-use efficiency   111

Good agricultural practices (GAP)   112

Organic soil amendments   113

Developing a circular economy   114

Economics of agriculture   115

Controlled environment agriculture   116

Understanding and conserving pollinators   117

Land use change and management   118

Understanding and tackling Fusarium wilt of banana   119

Novel biocontrol agents   120

Soil carbon sequestration   121

Soil contaminants   122

Environmental impact of poultry production   123

Silvopastoral systems   124

Acknowledgements

Chapters in this Instant Insight are taken from the following sources:

Chapter 1 Developing seaweed/macroalgae as feed for pigs
Chapter taken from: Chapter taken from: Lei, X. G. (ed.), Seaweed and microalgae as alternative sources of protein, Burleigh Dodds Science Publishing, Cambridge, UK, 2021, (ISBN: 978 1 78676 620 5; www.bdspublishing.com)

Chapter 2 High protein corn fermentation products for swine derived from corn ethanol production
Wiseman, J. (ed.), Advances in pig nutrition, Burleigh Dodds Science Publishing, Cambridge, UK, 2024, (ISBN: 978 1 80146 694 3; www. bdspublishing.com)

Chapter 3 Developing alternative sources of protein in pig nutrition: insects
Wiseman, J. (ed.), Advances in pig nutrition, Burleigh Dodds Science Publishing, Cambridge, UK, 2024, (ISBN: 978 1 80146 694 3; www. bdspublishing.com)

Chapter 4 Black soldier fly meal: an alternative protein source for pigs
Chapter taken from: Casillas, A., Insects as alternative sources of protein for food and feed pp. 3-28, Burleigh Dodds Science Publishing, Cambridge, UK, 2025, (ISBN: 978 1 80146 584 7; www.bdspublishing.com)

Chapter 1

Developing seaweed/macroalgae as feed for pigs

Marta López-Alonso, Universidade de Santiago de Compostela, Spain; Marco García-Vaquero, University College Dublin, Ireland; and Marta Miranda, Universidade de Santiago de Compostela, Spain

1 Introduction

2 Challenges in using macroalgae for feed applications

3 Composition of macroalgae

4 Biological functions and health-promoting effects of macroalgae and macroalgal-derived extracts in pig nutrition

5 Conclusion and future trends

6 Where to look for further information

7 Acknowledgements

8 References

1 Introduction

The global human population is projected to reach 9.7 billion by 2050. It has been estimated that overall food production will have to double by 2050 to meet the increased demand (Aiking, 2014). Animal farming systems will have to increase production by an estimated 70%, increasing the demand for animal feed by up to 235% (Herman and Schmidt, 2016). In meeting the increasing demand, intensive production systems have traditionally focused on maximising production outputs while minimising production costs (Jones et al., 2012). High-intensity farming has been characterised by genetically selecting animals for fast growth and high meat content, kept at high stocking densities and fed nutrient-dense plant-based feeds (Swanson, 1995). Potential consequences have been the greater levels of stress for animals and their vulnerability to disease, leading to the increased prophylactic use of antibiotics in some cases (Van Boeckel et al., 2015).

It has been increasingly recognised that some aspects of these production systems are no longer sustainable. Researchers have begun to investigate alternative feed sources that can be more sustainably produced than

http://dx.doi.org/10.19103/AS.2021.0091.15

conventional, land-based crops used for feed, and which also have the potential to boost animals' natural immunity to disease whilst meeting their nutritional requirements, thus reducing the reliance on antibiotics (García-Vaquero, 2018; Miranda et al., 2017; Morais et al., 2020).

Macroalgae or seaweeds comprise a diverse group with more than 10 000 different species described to date (Collins et al., 2016). Approximately 5% of them are currently exploited for feed applications (Michalak and Chojnacka, 2014). They are able to adapt to changing and extreme marine environmental conditions by producing secondary metabolites including lipids, polysaccharides and minerals, to protect the biomass from cell damage caused by these marine stressors (Garcia-Vaquero et al., 2017). These metabolites make them a promising potential feed ingredient.

The use of macroalgae as a source of protein and other essential nutrients (such as polyunsaturated fatty acids, phenolic compounds and minerals) in animal nutrition are promising dietary strategies to complement traditional feed sources (Øverland et al., 2019). The incorporation of macroalgal extracts containing carbohydrates such as laminarin and fucoidan has, for example, shown promising results as an alternative to antibiotics in piglets, either directly by feed supplementation and/or indirectly via maternal dietary supplementation (Sweeney and O'Doherty, 2016). This chapter summarises the nutritional properties of macroalgae for use in animal feed, both as a source of protein and as antimicrobial compounds in swine production systems.

2 Challenges in using macroalgae for feed applications

The use of macroalgae to feed livestock has been documented for thousands of years since at least the time of Ancient Greece (Makkar et al., 2016). In the nineteenth and early twentieth centuries, seaweed was used in the coastal regions of France, the Scottish islands and Scandinavia, mainly to feed ruminants but also other livestock species, including pigs (Chapman, 2012; Sauvageau, 1920). The use of brown algae like *Ascophyllum nodosum* or *Fucus vesiculosus* in pig nutrition has been documented since the beginning of the twentieth century. Macroalgae were traditionally mixed with cereal meals to fatten pigs in countries in Northern Europe (Chapman, 2012; Evans and Critchley, 2014). The interest in this feedstock increased during periods of food shortage such as World War I, but usage declined in the first half of the twentieth century due to the perceived poor nutritional quality of the biomass (Makkar et al., 2016). The low levels of protein and energy available for metabolism, and the high mineral content of brown seaweeds led to their replacement by other protein ingredients such as fishmeal and soybean meal in formulated feed for monogastric animals (Øverland et al., 2019). It was also found that high amounts of brown seaweed could be harmful to health: the inclusion of 10% of *A. Nodosum* in pig feed over several weeks, for example, has been related to weight loss (Jones et al., 1979). To maximise the potential benefits of the inclusion of seaweeds in feeds for improved pig production and health, seaweeds can only be used at low concentrations in feed (1–2%) rather than as a major source of macronutrients (Morais et al., 2020).

As previously mentioned, the high amount of ash in macroalgae has hindered the inclusion of intact or dried macroalgae in animal diets, as feeds high in minerals have resulted in diarrhoea and decreased animal performance in pigs and poultry (Koreleski et al., 2010; Wilkinson, 1992). There has been a recent focus on refining biomass to reduce unwanted mineral content in favour of the more beneficial compounds.

When extracting polysaccharides, their molecular weight, monosaccharide composition and sulphate contents will change depending on the extraction and purification methods used, affecting the biological properties and effects of these molecules (Garcia-Vaquero et al., 2017). Ale and Meyer (2013) attributed the lack of official approval of polysaccharides or their derived fractions for pharmaceutical, dermatological, nutraceutical or other commercial applications to the lack of standardised extraction methodologies.

Recent research has focused on identifying the optimum extraction parameters such as temperature, time and pH (Garcia-Vaquero et al., 2019; Yuan et al., 2018). Yuan et al. (2018) extracted sulphated polysaccharides from the green macroalga *Ulva prolifera* using microwave-assisted hydrothermal extraction. The sulphur content increased with the temperature and acid concentration of the solvent. Antioxidant and pancreatic lipase inhibition properties were also influenced by extraction conditions due to structural modifications of the active molecules (Yuan et al., 2018). Garcia-Vaquero et al. (2019) used hydrothermal-assisted extraction and found significant variation in the levels of fucoidan, laminarin and antioxidant activity under differing conditions. The authors determined the optimum process parameters for a combined maximised extraction of these compounds from *Laminaria hyperborea* (120°C, 80.9 min and 12.02 volume of solvent-to-seaweed ratio). Other techniques, such as the combination of ultrasound and microwave technologies, have also been explored for brown macroalga *A. nodosum*, achieving increased yields of fucoidan (Garcia-Vaquero et al., 2020).

Techniques such as enzyme-assisted extraction have also attracted attention due to their high efficiency and mild extraction conditions that ensure less modification in the chemical structure of the compounds of interest (Michalak and Chojnacka, 2014). Enzymes have been used to extract multiple compounds including proteins, phenols, carotenoids and lipids (Billakanti et al., 2013; Wang et al., 2010). However, the use of these methods has been hindered by the high price of enzymes and the cost of scaling up production (Michalak and Chojnacka, 2014).

Improvements in the extraction of beneficial compounds have been one factor in renewed scientific interest in macroalgae as a source of multiple high-value compounds including proteins, polysaccharides, lipids, pigments and minerals with a wide variety of antioxidant, anti-bacterial, anti-tumour and other properties (Holdt and Kraan, 2011). This has made them promising ingredients for a wide variety of applications, for example, as nutraceuticals or functional foods, pharmaceuticals, cosmetics and in animal feed.

3 Composition of macroalgae

Based on their pigment composition, macroalgae can be classified as:

- brown macroalgae (Phaeophyta);
- red macroalgae (Rhodophyta); and
- green macroalgae (Chlorophyta).

Macroalgae are rich in carbohydrates, with medium or high amounts of proteins, low levels of lipids and variable mineral content (Dominguez and Loret, 2019; Kraan, 2013). The composition of macroalgae varies greatly depending on the macroalgal class, species, parts of the macroalgae sampled as well as the season and location (Øverland et al., 2019). The variable composition of macroalgae (water, ash, crude protein, crude lipids and polysaccharides) compiled from the recent scientific literature are summarised in Table 1.

As seen in Table 1, the moisture content of the macroalgal biomass is high, accounting in some cases for up to 94% of the weight of the collected fresh biomass. This means that once the biomass is collected, the common process is drying (freeze-drying or oven drying) to preserve the biomass, before applying further processing or technological treatments to obtain macroalgal

Table 1 Summary of the composition ranges described for Phaeophyta (brown macroalgae), Chlorophyta (green macroalgae) and Rhodophyta (red macroalgae) in the recent scientific literature. The content of this table was modified from the work of Øverland et al. (2019)

Proximate composition	Phaeophyta	Chlorophyta	Rhodophyta
Water (% water in wet biomass)	61-94	78-92	72-91
Ash (% DW basis)	15-45	11-55	12-42.2
Polysaccharides (% DW basis)	38-61	15-65	36-66
Crude protein (% DW basis)	2.4-16.8	3.2-35.2	6.4-37.6
Crude lipids (% DW basis)	0.3-9.6	0.3-2.8	0.2-12.9

DW, dry weight.

ingredients (Garcia-Vaquero et al., 2020; Rioux et al., 2010; Yuan et al., 2018). The following sections review in more detail the composition of macroalgae: protein, carbohydrates, lipids, minerals and other compounds. This is partly because variations in these compounds are a key challenge in the effective use of macroalgae in feed, and partly because their accurate characterisation is essential to the use of bioactive compounds with beneficial effects in pig nutrition.

3.1 Protein content in macroalgae

Protein content varies significantly, depending on multiple factors including species. Comparing protein accumulation between studies is difficult due to the different approaches used to estimate protein content in macroalgae. The most commonly used method is based on the quantification of the overall nitrogen content, estimating the value of crude protein by using multiple conversion factors. Due to the high amounts of non-protein nitrogen in macroalgae, several conversion factors have been calculated (Biancarosa et al., 2017; Makkar et al., 2016). The conversion factor used in each study affects the comparisons between studies.

In general, protein concentrations in brown macroalgae are lower (1-16%) compared to green (11-26%) and red (11-33%) macroalgae that have protein levels comparable to other traditionally used protein-rich products (Garcia-Vaquero and Hayes, 2016). The protein concentration is also influenced by season. The average protein content of *Laminaria digitata*, for example, is 6.8%, with the highest levels accumulated during the first quarter of the year (Schiener et al., 2015). Interspecies and intraspecies variations in protein content are shown in Table 2 (García-Vaquero, 2018). This table summarises the variable protein contents (1.1-26.7%) of selected brown, red and green macroalgae.

As shown in Table 2, the reported protein content of brown macroalgae varied between 1.1% (Nielsen et al., 2016) and 16.1% (Mols-Mortensen et al., 2017). Huge variations in the protein content have also been reported for red macroalgal species, with levels varying from 9.2 % (Chan and Matanjun, 2017) to 25.2% (Lozano et al., 2016). The protein levels described in green macroalgae varied between 11.2 % (Biancarosa et al., 2017) and 26.7% (Silva et al., 2015). A significant variation in the levels of protein has also been described within these macroalgal groups depending on the species and climatological conditions affecting the biomass. Studying compounds in European kelp varieties (*Saccharina latissima* and *L digitata*) collected in Denmark, Nielsen et al. (2016) reported the highest biomass production and protein content in seaweed from highly saline water, while low salinity resulted in higher amounts of fermentable sugars and pigments. Manns et al. (2017) reported that the variation in composition was mainly related to season, species and location. The

Table 2 Protein content in selected brown, red and green macroalgae species, originally published by García-Vaquero (2018) and reproduced with permission from Springer Nature

Macroalgae (sp.)	Protein content (% dry weight)	References
Phaeophyta (brown macroalgae)		
Fucus sp.	3.9-4.0	(Biancarosa et al., 2017)
Ascophyllum nodosum	3.0	(Biancarosa et al., 2017)
	6.8	(Schiener et al., 2015)
Laminaria sp.	1.5-5.1	(Nielsen et al., 2016)
	6.6	(Biancarosa et al., 2017)
	2.4-12.7	(Manns et al., 2017)
Saccharina latissima	1.1-7.5	(Nielsen et al., 2016)
	4.8-14.1	(Manns et al., 2017)
	4.0-16.1	(Mols-Mortensen et al., 2017)
	10.6	(Stévant et al., 2017)
Eisenia arborea	9.2-12.5	(Landa-Cansigno et al., 2017)
Alaria esculenta	10.6	(Stévant et al., 2017)
Rhodophyta (red macroalgae)		
Porphyra sp.	13.5-20.6	(Biancarosa et al., 2017)
Gracilaria sp.	9.2	(Chan and Matanjun, 2017)
	12.6	(Chan and Matanjun, 2017)
Chondrus crispus	11.0	(Biancarosa et al., 2017)
Palmaria palmata	10.6	(Biancarosa et al., 2017)
	10.9	(Schiener et al., 2017)
Pyropia columbina	25.2	(Lozano et al., 2016)
Chlorophyta (green macroalgae)		
Ulva sp.	26.7	(Silva et al., 2015)
	13.6	(Angell et al., 2017)
	11.2-15	(Biancarosa et al., 2017)
	22.5	(Bikker et al., 2016)
Cladophora rupestris	12.0	(Biancarosa et al., 2017)

environmental variables studied by the authors included temperature, salinity, phosphate, nitrate and ammonia. The authors reported high levels of protein (15-20% w/w) in seaweed samples collected during February and March from multiple locations, while the lowest amount of protein (2.3-8.4% w/w) was reported in the biomass collected during July and August (Manns et al., 2017). Research on *Saccharina* spp. from the Faroe Islands by Mols-Mortensen et al. (2017) found significantly higher protein content in biomass collected in April-May compared to July-August, while no significant differences were found between locations (Mols-Mortensen et al., 2017).

In contrast to carbohydrates, protein levels are highest during the winter and low during the summer (Adams et al., 2011; Fleurence, 1999; Schiener et al., 2015). It has been suggested that high levels of protein support the build-up of nitrogen reserves for rapid growth during the summer months (Chapman and Craigie, 1977). Schiener et al. (2015) reported the average contents of protein at 6.9 ± 1.1%, 6.8 ± 1.3%, 7.1 ± 1.7% and 11.0 ± 1.4% in *L. digitata, L. hyperborea, S. latissima* and *A. esculenta*, respectively. The contents of protein varied significantly through the year, with high levels of protein during the first quarter of the year for all macroalgal species, while the lowest protein contents were found in the third quarter of the year.

Amino acids combine to make proteins. Glutamic and aspartic acids are the most abundant amino acids in most macroalgae, with generally low levels of methionine (Holdt and Kraan, 2011). Brown macroalgae are considered rich sources of threonine, valine, leucine, lysine, glycine and alanine (Holdt and Kraan, 2011). Macroalgae usually contain high levels of glutamic acid, present both in its bound and free forms, contributing to the characteristic umami taste of algae (Mæhre et al., 2014). Lourenço et al. (2002) studied the amino acid and protein composition of 19 tropical seaweed species. The authors reported that green algae had low percentages of aspartic and glutamic acids, while red algae showed high percentages of lysine and arginine, and brown macroalgae had higher amounts of methionine. The authors also reported significant variations in the concentration of individual amino acids depending on macroalgal species. The highest concentration of glutamic acid (17.6% of total amino acids) was found in the brown macroalga *Sargassum vulgare*, while the lowest amount (10.7%) was reported in the green macroalga *Caulerpa fastigiate*. The amino acid content varied between species, location and season of collection. Manns et al. (2017) reported that the brown macroalgae *L. digitata* and *S. latissima* contained mainly glutamic acid, aspartic acid and alanine. Individual amino acid contents vary significantly between spring and summer, with levels more dependent on location than species. Glutamic acid represented over 25% of the total amino acid content of *L. digitata* collected in March, while these levels decreased to 12.9% in biomass collected in August. The highest levels of aspartic acid were found in August (14.5%) compared to March (9.5%).

The high levels of protein in macroalgae make it possible to produce bioactive peptides or cryptides. Bioactive peptides are sequences of 2–30 amino acids with no activity within the parent protein. After their release by several enzymatic hydrolysis or fermentation processes, these peptides have been linked to a wide range of health benefits in food and feed. Several bioactive peptides have been identified from *Undaria pinnatifida* (Suetsuna and Nakano, 2000). Some have been commercialised, such as the Wakame peptide jelly (Riken Vitamin Co., Ltd., Tokyo, Japan) and Nori peptide S (Shirako Co., Ltd., Tokyo, Japan) (Fukami, 2010).

3.2 Carbohydrates in macroalgae

Carbohydrates are one of the major components of macroalgae, with contents ranging from 4% to 76% depending on species, location and season (Holdt and Kraan, 2011). Macroalgal carbohydrates include a wide range of compounds (Lafarga et al., 2020). The major polysaccharides include alginate, carrageenan, and other phycocolloids such as agar, commonly used as stabilisers, thickeners and emulsifiers in food production. Red macroalgae mainly contain floridean starch, cellulose and other minor compounds such as mannans, xylans and sulphated galactans. Other carbohydrates, such as glucans (laminarin) and fucoidan, are mainly present in brown macroalgae, while green macroalgae are the main source of energy-reserve carbohydrates such as ulvans (Lafarga et al., 2020; Stiger-Pouvreau et al., 2016). As noted, the levels depend on factors such as species and season. Fucoidan levels in *F. vesiculosus,* for example, were higher than in *U. pinnatifida* (Holdt and Kraan, 2011). The highest amount of fucoidan in cultured *L. japonica* occurred when the biomass matured, rather than during the early stage of development (Honya et al., 1999).

Macroalgal carbohydrates such as fucoidan and laminarin have been associated with improved piglet performance and gut health (Gahan et al., 2009; McDonnell et al., 2010), and have been considered as alternatives to antibiotics and zinc oxide (Sweeney and O'Doherty, 2016). These polysaccharides demonstrate a range of anti-inflammatory, antioxidant, anti-coagulant, anti-viral and anti-tumour properties *in vitro* and *in vivo* (Garcia-Vaquero et al., 2017), with the potential to benefit human health through improving the antioxidant content of animal-derived products (Moroney et al., 2012, 2015). Dietary supplementation of extracts from brown seaweed *(L. digitata)* containing laminarin and fucoidan also improved the quality and shelf life of pig meat by lowering lipid oxidation (Moroney et al., 2012, 2015). Since polysaccharides such as alginates, fucoidan, and laminarin are not digested in the upper digestive tract of the animals, they can be considered as a source of dietary fibre for animals (Garcia-Vaquero et al., 2017). Lahaye (1991) estimated total dietary fibre ranging from 32.7% to 74.6% (on a DW basis) and water-soluble fractions ranging from 51.6% to 85%.

3.3 Lipids and other compounds in macroalgae

In general, macroalgae have a low lipid content, normally ranging between 0.2% and 12.9% as seen in Table 1. However, macroalgal lipids are rich in polyunsaturated fatty acids (PUFA), such as omega 6 and 3 (n6 and n3), with health benefits for both humans and animals. The relative abundance of a particular PUFA depends on the species. Red macroalgae normally have high concentrations of eicosapentaenoic acid (C20:5, n3) that can reach levels of up to half of the total fatty acids of the macroalgae; while brown macroalgae have in general high concentrations of oleic acid (C18:1, n9), linoleic acid (C18:2, n6) and α-linolenic acid (C18:3, n3) (Dawczynski et al., 2007).

Macroalgae are also rich in other phytochemicals including phenolic compounds, pigments and vitamins. As previously mentioned, the main macroalgal classes are established on the basis of the pigmentation of the biomass. Brown macroalgae are rich in fucoxanthin, while the colour of red macroalgae is associated with an abundance of phycobilins and other pigments such as chlorophylls (a and b) and carotenes. Xanthophylls are the main compounds influencing the pigmentation of green macroalgae (Øverland et al., 2019). Macroalgae also contain a wide range of phenolic compounds (i.e. phlorotannins and phloroglucinols) with potent antioxidant properties (Rajauria, 2018). Roleda et al. (2019) reported differences in phenolic compound levels depending on the species and season of collection. The levels of polyphenols in *Palmaria palmata* were significantly higher in Spring, compared to Summer and Autumn (Roleda et al., 2019).

3.4 Mineral contents in macroalgae

The amount of ash in the macroalgal biomass, ranging from 11% to 55% (Table 1), hinders the utilisation of full or intact dried macroalgae as animal feed. The incorporation of high levels of minerals in the feed of monogastric species, including pigs and poultry, resulted in diarrhoea and decreased animal performance (Koreleski et al., 2010; Wilkinson, 1992). There has been a recent study on the extraction of high-value compounds, concentrating them while decreasing the amount of minerals present in the raw biomass. Several innovative technologies, including ultrasound and microwaves, have been explored to obtain macroalgal ingredients or extracts rich in biologically active compounds, such as polysaccharides (i.e. fucoidan and laminarin), for feed applications (Garcia-Vaquero et al., 2017; Yuan et al., 2018).

Macroalgae can accumulate variable levels of minerals including essential trace elements (Ca, Co, Cr, Cu, Fe, I, Mg, Mn, Mo, Ni, P, Se and Zn), providing a source of essential minerals in animal feed (Rey-Crespo et al., 2014). The mineral levels vary by species due to varying metal-binding capacities (Güven et al., 1995). The cell wall polysaccharides produced by brown macroalgae have a high ability to absorb and retain metals. As a result, alginates produced by brown macroalgae have the strongest metal-binding capacity, followed by carrageenans and agar, predominantly produced by red macroalgae (Güven et al., 1995). Brown macroalgae have been described as being exceptionally high in iodine, compared to red and green macroalgae (Biancarosa et al., 2018; Mæhre et al., 2014).

Biancarosa et al. (2018) analysed the mineral profile of 21 macroalgal species collected from the coasts of Norway, and reported generally high iodine contents, varying from 22 mg/kg to 10 000 mg/kg dried weight (DW), with the highest accumulation reported in the brown macroalgae *L digitata*. Seaweed supplements in pigs are considered a way to increase the iodine content of meat (Dierick et al., 2009; He et al., 2002). Dierick et al. (2009) demonstrated that feeding pigs with 2% of *A. nodosum* increased the concentration of iodine in the tissue by 2.7 to 6.8, depending on the tissue.

Macroalgae are also known to accumulate toxic metals such as As, Cd and Hg. Yamada et al. (2007) reported different concentrations of Cd and Pb in *U pinnatifida* related to industrial waste. A number of regulators are addressing content of toxic metals in macroalgae when used as animal feed. In the European Union, calcareous marine algae must contain less than 10 mg As/kg and 15 mg Pb/kg relative to a feed, with moisture content of 12% (Commission Regulation 574/2011) (Regulation, 2011). A feedstuff containing macroalgae must contain less than 40 mg As/kg, with levels of the most toxic As species (inorganic As) of less than 2 mg/kg (Commission Regulation 574/2011) (Regulation, 2011). European legislation specifically mentions levels of As in the brown macroalga *Hizikia fusiforme*, as a result of previous reports of high levels (Besada et al., 2009; Rose et al., 2007).

Most studies analysing the mineral content of macroalgae have focused on analysing total As concentrations (Biancarosa et al., 2018; Khan et al., 2015; Ronan et al., 2017), even though As may be present in its organic form with little or no toxicity (Contam, 2009). Rose et al. (2007) reported that the level of inorganic As was less than 7% of total As. A recent study determining the mineral profile of three brown macroalgae (*L digitata, L hyperborea* and *A. nodosum*) collected in Ireland reported low levels of toxic metals (Cd, Hg and Pb), while levels of total As were high (49–64 mg/kg DW macroalgae) compared with previous reports (Garcia-Vaquero et al., 2021).

4 Biological functions and health-promoting effects of macroalgae and macroalgal-derived extracts in pig nutrition

Based on the properties of macroalgae, particularly brown seaweeds, a large body of recent research has been conducted in pigs. Table 3 summarises the experimental work conducted in the last decade on the use of macroalgae or macroalgal extracts in pigs.

There is limited information on the effects of dietary supplementation with full macroalgal biomass. Choi et al. (2017) described the beneficial effects on growth performance, gut microflora and intestinal morphology in weaning pigs receiving intact *Ecklonia cava* that could be attributed to the high levels of fucoidan (ca. 112 ± 6 g/Kg) in this brown macroalgae. However, Michiels et al. (2012) reported that the addition of *A. nodosum* had no effect on the

Table 3 Effects of macroalgae supplementation on the performance and gut health in weaned piglets. Supplemented diets were tested against a control diet

Compounds/Substances tested (doses)	Duration (days)	End-points*						References
		Growth performance	Nutrient digestibility	Gut microbiota	Gut architecture	Antioxidant activity	Immune system modulation	
Macroalgal extracts								
Laminarin (300 mg/kg) and/or fucoidan (240 mg/kg) extracts	Weaning to 8 d			+	+		+	(Walsh et al., 2013a)
Brown seaweed (alginates oligosaccharides (50–200 mg/kg)	Weaning to 14 d	+	+			+	+	(Wan et al., 2017)
Brown seaweed (alginic acid oligosaccharides (100 mg/kg)	Weaning to 14 d				+	+	+	(Wan et al., 2018)
Fucoidan extract (44%), (125 and 250 mg/kg)	Weaning to 14 d	-	+					(Rattigan et al., 2019)
Laminarin extract (65%) (100, 200 and 300 mg/kg)	Weaning to 14 d	+		+				(Rattigan et al., 2020)
Laminarin extract (65%) (300 mg/kg)	Weaning to 14 d	+		+	+			(Vigors et al., 2020)
Brown seaweed (alginic acid oligosaccharides (100 mg/kg)	Weaning to 21 d	+	+	+	+	+	+	(Wan et al., 2016)
Laminarin (300 mg/kg) and/or fucoidan (240 mg/kg) extracts	Weaning to 21 d	+		+				(McDonnell et al., 2010)

(Continued)

Table 3 (*Continued*)

Compounds/Substances tested (doses)	Duration (days)	End-points*						References
		Growth performance	Nutrient digestibility	Gut microbiota	Gut architecture	Antioxidant activity	Immune system modulation	
Laminarin (300 mg/kg) and/or fucoidan (236 mg/kg) extracts	Weaning to 21 d	+		+				(O'Doherty et al., 2010)
Laminarin (300 mg/kg) and/or fucoidan (240 mg/kg) extracts	Weaning to 21 d	+	+					(McAlpine et al., 2012)
Seaweed extract containing laminarin (314 mg/kg) and fucoidan (249 mg/kg)	Weaning to 25 d	+	+	+				(Dillon et al., 2010)
Laminarin (300 mg/kg) and/or fucoidan (240 mg/kg) extracts	Weaning to 32 d	+	+	+				(Heim et al., 2014)
Laminarin (150–300 mg/kg) and/or fucoidan (240 mg/kg) extracts	Weaning to 35 d	+	+	+				(Walsh et al., 2013b)
Fucoidan extract (44%), (250 mg/kg) and laminarin extract (65%) (300 mg/kg)	Weaning to 35 d	-	+	+	+			(Vigors et al., 2021)
Laminarin (300 mg/kg) and/or fucoidan (240 mg/kg) extracts	Weaning to 40 d	+	+	+				(O'Shea et al., 2014)

Treatment	Inclusion period						Reference
Commercial seaweed extract (OceanFeed Swine®) (5 g/kg)	Weaning to 160 d	+		+			(Ruiz et al., 2018)
Seaweed extract containing laminarin (sows 1 g/d; piglets 300 mg/kg) and fucoidan (sows 0.8 g/d; piglets 240 mg/kg)	107 d gestation-weaning (28 d)	+		+			(Draper et al., 2016)
Seaweed extract (10 g/d) containing laminarin (1 g) and fucoidan (0.8 g)	107 d gestation-weaning (28 d)	+			+	+	(Heim et al., 2015a)
Seaweed extract containing laminarin (280 mg/kg) and fucoidan (244 mg/kg)	107 d gestation-weaning (26 d)	+	–	+	–	+	(Leonard et al., 2011)
Seaweed extract (10 g/d) containing laminarin (1 g) and/or fucoidan (0.8 g)	107 d gestation-weaning (24 d)	+			+	+	(Heim et al., 2015b)
Whole macroalgae							
Ecklonia cava (0.5–1.5 g/kg)	Weaning to 28 d	+	–	+	+	+	(Choi et al., 2017)
Ascophyllum nodosum (2.5–10 g/kg)	Weaning to 28 d	–		–			(Michiels et al., 2012)

*Key of outputs: + (significant positive); – (significant negative) effect of the treatment compared to the control group; empty cell, not indicated or not determined.

performance or gut health and plasma oxidative status of piglets, possibly due to levels of bioactive compounds being too low to exert any prebiotic effect.

As shown in Table 3, many studies involve the use of macroalgal extracts in feed. The extracts studied mainly include mixtures of carbohydrates, such as laminarin and/or fucoidan, given as feed supplements to piglets during the post-weaning period (from weaning up to 40 days), since the weaning phase is a critical period due to the high incidence of enteric pathologies. Supplements have also been given to sows at the end of the gestation period, and the effects of this maternal supplementation on piglets were then analysed (Heim et al., 2015a,b; Leonard et al., 2011). Other studies have focused on the long-term effects of dietary supplementation fed to animals from weaning to slaughter (Draper et al., 2016; Ruiz et al., 2018).

4.1 Growth performance and digestion

The studies shown in Table 3 reported the overall positive effects on productive performance with the inclusion of macroalgal extracts in the diet. The average daily gain (ADG) of piglets fed brown seaweeds and/or macroalgal-derived ingredients was in all cases higher (ca. 5% to 40%) compared to that of piglets fed a control diet. Studies evaluating animals at slaughter found the effects of supplementation on ADG to be limited, with improvements of < 5 % compared to those receiving a control diet (Draper et al., 2016; Heim et al., 2015b). The positive effects of macroalgal extracts on growth seem to be related to the prebiotic effects of seaweed polysaccharides on piglet gut function and health (Corino et al., 2019). Overall, the inclusion of macroalgal extracts had positive effects on the digestibility of nitrogen (up to 8% increase), gross energy (up to 10% increase), fibre (up to 73% increase) and ash (up to 80% increase) in most of the studies. The improvement in nutrient digestibility seems to be related to the effect of carbohydrates and antioxidants on gut microbiota and on the villous architecture, with an increase in absorptive capacity and nutrient transporters (Sweeney and O'Doherty, 2016), and also to the beneficial effect on the intestinal mucosal cells and volatile fatty acid production (i.e. butyric acid) (Corino et al., 2019).

4.2 Gut function

Macroalgal supplements also have significant effects on gut microbiota (Table 3). Algal supplements stimulate the growth of *Lactobacilli* and reduce the population of Enterobacteriaceae such as *Escherichia coli*. Dierick et al. (2010) observed that 1% of *A. nodosum* added to feed reduced *E. coli*, while increasing the Lactobacilli/*E. coli* ratio, leading to a lower susceptibility of the animals to intestinal disorders.

Several studies have also demonstrated the positive influence of extracts from brown seaweed on the morphological features of the small intestine of suckling piglets, with increased villus height in the ileum. These positive effects have been attributed to an upregulation in the expression of the tight junction proteins, occludin (OCLN) and zonula occludens 1 (ZO-1) in the small intestine (Wan et al., 2016), together with enhanced antioxidant capacity, decreased mast cell degranulation and prevention of the release of pro-inflammatory cytokines via restraining the TLR4/NF-kB and NOD1/NF-kB signalling pathways (Wan et al., 2016, 2017, 2018). Maternal dietary supplementation with macroalgal-derived polysaccharides also down-regulated the gene expression of pro-inflammatory cytokines involved in *E. coli*-caused diarrhoea in piglets at 48 h after birth and weaning (Heim et al., 2015a).

A number of recent studies have focused on the effects of feeding fucoidan and laminarin for 14 days post-weaning (Rattigan et al., 2019; Vigors et al., 2020). Rattigan et al. (2020) found that laminarin supplementation at 300 ppm provided the most beneficial effects, with improvements in animal performance and positive effects in small intestinal morphology, microbial populations and gene expression. When analysing the microbiome of the laminarin-supplemented group, the changes observed included an increased abundance of *Prevotella* and reductions in *Enterobacteriaceae* (Vigors et al., 2020). Laminarin supplementation had a positive influence on intestinal health through alterations in the gastrointestinal microbiome and increased butyrate production (Vigors et al., 2021). However, there was no additional benefit on performance with either laminarin or fucoidan supplementation up to day 35 post-weaning.

4.3 Immune function and antioxidant capacity

Seaweed extracts also seem to play an immunomodulatory role. Leonard et al. (2011) reported that the dietary supplementation of sows from 109 days of gestation until weaning with extracts from the brown macroalga *Laminaria spp.* increased levels of immunoglobulins (Ig) G and A in the colostrum of sows by 19% and 25%, respectively, as well as the levels of IgG in the serum of piglets by 10%. Feeding seaweed extracts during lactation suppressed pro-inflammatory IL-1a mRNA expression in the ileum of pigs 11 days after weaning. Dietary seaweed extract supplementation post-weaning also induced up-regulation in colonic MUC2 mRNA expression in pigs 11 days post-weaning. These results demonstrate that feed supplementation post-weaning can improve gut health and growth performance in starter pigs.

Other macroalgal polysaccharides such as alginic acid supplemented at 100 mg/kg feed over 21 days post-weaning also increased IgG and IgA concentrations in piglet serum by 20% and 53%, respectively (Wan et al., 2016).

Alginic acid supplementation also increased superoxide dismutase, catalase and total antioxidant activities, and decreased the concentration of malonic dialdehyde in the serum of supplemented animals. The positive effects of alginic acid on the growth performance, antioxidant capacity, immunity and intestinal development in weaned pigs suggest that alginic acid could serve as a useful bioactive feed additive.

Other studies have evaluated the effect of dietary supplementation with laminarin and fucoidan on antioxidant activity (Moroney et al., 2012, 2015; Rajauria et al., 2016). Dietary supplementation of laminarin and fucoidan (0.45-0.9 g/kg) in pigs 3 weeks pre-slaughter resulted in enhanced meat due to deposition of marine-derived bioactive antioxidant components in muscle, reduced saturated fatty acids and lowered lipid oxidation in muscle (Moroney et al., 2015). This antioxidant response was attributed to mechanisms such as the reduction of saturated fatty acids and decreased lipid oxidation in the muscle, though it is still unclear if the free radical scavenging abilities of the extract are responsible for the antioxidant activity observed in the muscle. The enhanced meat quality effected by seaweed polysaccharides may be mediated through the health-promoting effects of gut-associated immunity. The improved fatty acid profiles and enhanced lipid stability of pork meat had no impact on the tenderness, flavour or other sensory properties of the meat. This suggests that dietary supplementation of seaweed extracts containing laminarin and fucoidan in pigs could result in an enhanced meat product. Rajauria et al. (2016) found an improvement in meat quality from the addition of a seaweed extract containing laminarin, due to decreased lipid oxidation. Improved antioxidant capacity and colour stability, together with the reduced lipid peroxidation in the muscle longissimus dorsi, suggest that feed supplementation with macroalgal polysaccharides may protect animals from oxidative stress-induced diseases and improve meat quality.

5 Conclusion and future trends

So far, beneficial effects have been mainly found when adding macroalgae at levels of ≤ 10% of the total concentration in animal feed. More studies of the biochemical profile of seaweed macro and micronutrients, and particularly seaweed bioactive metabolites are needed to properly calculate optimum rates of inclusion of the macroalgae in feed, both to optimise the beneficial effects and to avoid the negative or toxic effects of other compounds (i.e. As, Cd and Hg toxic metals). Wild seaweed biomass varies enormously in nutritional value because of seasonal geographical variations and varying risks of bioaccumulation of heavy metals. Regular monitoring is essential to provide a reliable source of safe animal feed supplementation. A large body of research has been conducted in the last decade on the supplementation of pigs' feed with macroalgal extracts, with beneficial effects related mainly to the presence of polysaccharides (laminarin and/or fucoidan) extracted from brown macroalgae. These studies demonstrate the potential of macroalgal compounds as feed supplements and their application in pig nutrition and production. Despite these promising results, particularly when used as substitutes for in-feed antibiotics, further research is needed to elucidate the effects of individual compounds present in macroalgal extracts. The production of extracts following optimised and standardised protocols, and the characterisation of the molecules within these extracts, will allow the establishment of clear structure–function relationships. This will allow industrial exploitation of macroalgal extracts for improved intestinal physiology, morphology, microbiology and immune response of post-weaning piglets, leading to industrial products with proven health-enhancement benefits that can be applied routinely in the herds.

6 Where to look for further information

As seen in the current chapter, macroalgae can be an excellent source of multiple compounds that can be used in animal feed for multiple purposes. However, the incorporation of macroalgae in animal feed will ultimately depend on the composition of the original feedstock. An introduction to the wide variation of the macroalgal biomass with respect to the species, season and year of collection can be accessed in Garcia-Vaquero et al. (2021). 'Seasonal variation of the proximate composition, mineral content, fatty acid profiles and other phytochemical constituents of selected brown macroalgae' (see full details in the reference list). This publication may serve as a reference guide for the inclusion of macroalgae in animal feed and the selective collection of this feedstock depending on the intended use in animals' diet. Moreover, one of the main organisations in Europe promoting the knowledge on algal biomass is the European Algae Biomass Association (EABA) (https://www.eaba-ass ociation.org/en/aboutus). This group include key information related to algal biomass and main key stakeholders of this industry in Europe.

Macroalgae, despite being an excellent source of multiple beneficial compounds, may also contain high levels of minerals and other minor compounds (heavy metals) that may limit the utilisation of this biomass either for practical purposes or legal issues related to the presence of these contaminants in the biomass. An extensive legislative framework on the use of macroalgae as animal feed in Europe can be accessed from Garcia-Vaquero and Hayes (2016) 'Red and green macroalgae for fish and animal feed and human functional food development' (see in the reference list). Furthermore, the use of macroalgae in animal feed is regulated in Europe by the European Food Safety Authority (EFSA) within the 'feed additives' panel. The readers of the chapter are referred to the EFSA and, in particular, to the publications of this panel to get the latest information related to the use of any fed additive (https://www.efsa.europa.eu/en/topics/topic/feed-additives). Individual countries may also provide their own guidance or interpretation of these European guidelines. As an example, the Food Safety Authority of Ireland (FSAI) published the report 'Safety Considerations of Seaweed and Seaweed-derived Foods Available on the Irish Market' (ISBN: 978-1-910348-42-0) and established recommendations for the seaweed sector in Ireland in addition to the aforementioned European regulations.

As the field of animal nutrition specialises, further studies will be needed not only including macroalgae or macroalgal extracts in animal feed, but targeting specific macroalgal molecules in animal feed to establish a clear association between the health benefits appreciated in the animals and particular macroalgal compounds that could be industrially exploited. In order to achieve that, multidisciplinary projects, including the optimisation of the extraction of macroalgal compounds as well as their *in vitro* and *in vivo* evaluation, should be further explored. Currently, the European project BIOCARB-4-FOOD (17RDSUSFOOD2ERA-NET1, https://www.biocarb4food.eu/) focuses on targeting the extraction and utilisation of macroalgal carbohydrates for multiple industrial applications. Other promising projects following also a multidisciplinary approach include the 'Enhance microalgae' (https://www.enhancemicroalgae.eu/). This European project offers a comprehensive approach to the study of algae for the generation of high-value compounds, starting from cultivation to extraction of these compounds from algae using novel technologies, as well as the final potential applications of these compounds in multiple industries.

7 Acknowledgements

This chapter did not receive any specific grant from funding agencies in the public, commercial or not-for-profit sectors.

8 References

Adams, J. M. M., Toop, T. A., Donnison, I. S. and Gallagher, J. A. (2011). Seasonal variation in Laminaria digitata and its impact on biochemical conversion routes to biofuels. *Bioresource Technology* 102(21): 9976–9984. DOI: 10.1016/j.biortech.2011.08.032.

Aiking, H. (2014). Protein production: planet, profit, plus people? *The American Journal of Clinical Nutrition* 100(Suppl. 1): 483S–489S. DOI: 10.3945/ajcn.113.071209.

Ale, M. T. and Meyer, A. S. (2013). Fucoidans from brown seaweeds: an update on structures, extraction techniques and use of enzymes as tools for structural elucidation. *RSC Advances*. Royal Society of Chemistry 3(22): 8131–8141. DOI: 10.1039/C3RA23373A.

Angell, A. R., Paul, N. A. and de Nys, R. (2017). A comparison of protocols for isolating and concentrating protein from the green seaweed Ulva ohnoi. *Journal of Applied Phycology* 29(2): 1011–1026. DOI: 10.1007/s10811-016-0972-7.

Besada, V., Andrade, J. M., Schultze, F. and González, J. J. (2009). Heavy metals in edible seaweeds commercialised for human consumption. *Journal of Marine Systems* 75(1–2): 305-313. DOI: 10.1016/j.jmarsys.2008.10.010.

Biancarosa, I., Espe, M., Bruckner, C. G., Heesch, S., Liland, N., Waagbø, R., Torstensen, B. and Lock, E. J. (2017). Amino acid composition, protein content, and nitrogen-to-protein conversion factors of 21 seaweed species from Norwegian waters. *Journal of Applied Phycology* 29(2): 1001-1009. DOI: 10.1007/s10811-016-0984-3.

Biancarosa, I., Belghit, I., Bruckner, C. G., Liland, N. S., Waagbø, R., Amlund, H., Heesch, S. and Lock, E. J. (2018). Chemical characterization of 21 species of marine macroalgae common in Norwegian waters: benefits of and limitations to their potential use in food and feed. *Journal of the Science of Food and Agriculture*. John Wiley & Sons, Ltd 98(5): 2035-2042. DOI: 10.1002/jsfa.8798.

Bikker, P., van Krimpen, M. M., van Wikselaar, P., Houweling-Tan, B., Scaccia, N., van Hal, J. W., Huijgen, W. J. J., Cone, J. W. and López-Contreras, A. M. (2016). Biorefinery of the green seaweed Ulva Lactuca to produce animal feed, chemicals and biofuels. *Journal of Applied Phycology*. Springer Netherlands 28(6): 3511-3525. DOI: 10.1007/s10811-016-0842-3.

Billakanti, J. M., Catchpole, O. J., Fenton, T. A., Mitchell, K. A. and MacKenzie, A. D. (2013). Enzyme-assisted extraction of fucoxanthin and lipids containing polyunsaturated fatty acids from Undaria pinnatifida using dimethyl ether and ethanol. *Process Biochemistry* 48(12): 1999-2008. DOI: 10.1016/j.procbio.2013.09.015.

Chan, P. T. and Matanjun, P. (2017). Chemical composition and physicochemical properties of tropical red seaweed, Gracilaria changii. *Food Chemistry* 221: 302-310. DOI: 10.1016/j.foodchem.2016.10.066.

Chapman, A. R. O. and Craigie, J. S. (1977). Seasonal growth in Laminaria longicruris: relations with dissolved inorganic nutrients and internal reserves of nitrogen. *Marine Biology* 40(3): 197-205. DOI: 10.1007/BF00390875.

Chapman, V. (2012). *Seaweeds and Their Uses*. New York: Springer Science & Business Media.

Choi, Y., Hosseindoust, A., Goel, A., Choi, Y., Hosseindoust, A., Goel, A., Lee, S., Jha, P. K., Kwon, I. K., & Chae, B. J. (2017). Effects of Ecklonia cava as fucoidan-rich algae on growth performance, nutrient digestibility, intestinal morphology and caecal microflora in weanling pigs. *Asian-Australasian Journal of Animal Sciences*. Asian-Australasian Association of Animal Production Societies (AAAP) and Korean Society of Animal Science and Technology (KSAST) 30(1): 64-70. DOI: 10.5713/ajas.16.0102.

Collins, K. G., Fitzgerald, G. F., Stanton, C. and Ross, R. P. (2016). Looking beyond the terrestrial: the potential of seaweed derived bioactives to treat non-communicable diseases. *Marine Drugs*. MDPI AG 14(3). DOI: 10.3390/md14030060.

Contam EP on C in the FC (2009) Scientific opinion on arsenic in food. *EFSA Journal* Wiley-Blackwell Publishing Ltd 7(10). DOI: 10.2903/j.efsa.2009.1351.

Corino, C., Modina, S. C., Di Giancamillo, A., Chiapparini, S. and Rossi, R. (2019). Seaweeds in pig nutrition. *Animals: An Open Access Journal from MDPI*. MDPI AG 9(12): 1126. DOI: 10.3390/ani9121126.

Dawczynski, C., Schubert, R. and Jahreis, G. (2007). Amino acids, fatty acids, and dietary fibre in edible seaweed products. *Food Chemistry* 103(3): 891-899. DOI: 10.1016/j.foodchem.2006.09.041.

Dierick, N., Ovyn, A. and De Smet, S. (2009). Effect of feeding intact brown seaweed Ascophyllum nodosum on some digestive parameters and on iodine content in edible tissues in pigs. *Journal of the Science of Food and Agriculture* 89(4): 584-594. DOI: 10.1002/jsfa.3480.

Dierick, N., Ovyn, A. and De Smet, S. (2010). In vitro assessment of the effect of intact marine brown macro-algae Ascophyllum nodosum on the gut flora of piglets. *Livestock Science* 133(1-3): 154-156. DOI: 10.1016/j.livsci.2010.06.051.

Dillon, S., Sweeney, T., Figat, S., Callan, J. J. and O'Doherty, J. V. (2010). The effects of lactose inclusion and seaweed extract on performance, nutrient digestibility and microbial populations in newly weaned piglets. *Livestock Science* 134(1-3): 205-207. DOI: 10.1016/j.livsci.2010.06.142.

Dominguez, H. and Loret, E. P. (2019). Ulva Lactuca, A source of troubles and potential riches. *Marine Drugs*. MDPI AG 17(6): 357. DOI: 10.3390/md17060357.

Draper, J., Walsh, A. M., McDonnell, M. and O'Doherty, J. V. (2016). Maternally offered seaweed extracts improves the performance and health status of the postweaned pig1. *Journal of Animal Science* 94(suppl_3): 391-394. DOI: 10.2527/jas.2015-9776.

Evans, F. D. and Critchley, A. T. (2014). Seaweeds for animal production use. *Journal of Applied Phycology* 26(2): 891-899. DOI: 10.1007/s10811-013-0162-9.

Fleurence, J. (1999). Seaweed proteins. *Trends in Food Science and Technology* 10(1): 25-28. DOI: 10.1016/S0924-2244(99)00015-1.

Fukami, H. (2010). *Functional Foods and Biotechnology in Japan*. DOI: 10.1201/9781420087123.

Gahan, D. A., Lynch, M. B., Callan, J. J., O'Sullivan, J. T. and O'Doherty, J. V. (2009). Performance of weanling piglets offered low-, medium- or high-lactose diets supplemented with a seaweed extract from Laminaria spp. *Animal* 3(1): 24-31. DOI: 10.1017/S1751731108003017.

Garcia-Vaquero, M. and Hayes, M. (2016). Red and green macroalgae for fish and animal feed and human functional food development. *Food Reviews International*. Taylor & Francis 32(1): 15-45. DOI: 10.1080/87559129.2015.1041184.

Garcia-Vaquero, M., Rajauria, G., O'Doherty, J. V. and Sweeney, T. (2017). Polysaccharides from macroalgae: recent advances, innovative technologies and challenges in extraction and purification. *Food Research International* 99(3): 1011-1020. DOI: 10.1016/j.foodres.2016.11.016.

García-Vaquero, M. (2018). Seaweed proteins and applications in animal feed. In: *Novel Proteins for Food, Pharmaceuticals and Agriculture*. Wiley Online Books, pp. 139-161. DOI: 10.1002/9781119385332.ch7.

Garcia-Vaquero, M., O'Doherty, J. V., Tiwari, B. K., Sweeney, T. and Rajauria, G. (2019). Enhancing the extraction of polysaccharides and antioxidants from macroalgae using sequential hydrothermal-assisted extraction followed by ultrasound and thermal technologies. *Marine Drugs*. MDPI AG 17(8): 457. DOI: 10.3390/md17080457.

Garcia-Vaquero, M., Ummat, V., Tiwari, B. and Rajauria, G. (2020). Exploring ultrasound, microwave and ultrasound-microwave assisted extraction technologies to increase the extraction of bioactive compounds and antioxidants from brown macroalgae. *Marine Drugs*. MDPI AG 18(3): 172. DOI: 10.3390/md18030172.

Garcia-Vaquero, M., Rajauria, G., Miranda, M., Sweeney, T., Lopez-Alonso, M. and O'Doherty, J. (2021). Seasonal variation of the proximate composition, mineral content, fatty acid profiles and other phytochemical constituents of selected brown macroalgae. *Marine Drugs* 19(4). DOI: 10.3390/md19040204.

Güven, K. C., Akyüz, K. and Yurdun, T. (1995). Selectivity of heavy metal binding by algal polysaccharides. *Toxicological and Environmental Chemistry*. Taylor & Francis 47(1-2): 65-70. DOI: 10.1080/02772249509358127.

He, M. L., Hollwich, W. and Rambeck, W. A. (2002). Supplementation of algae to the diet of pigs: a new possibility to improve the iodine content in the meat. *Journal of Animal Physiology and Animal Nutrition* 86(3-4): 97-104. DOI: 10.1046/j.1439-0396.2002.00363.x.

Heim, G., Walsh, A. M., Sweeney, T., Doyle, D. N., O'Shea, C. J., Ryan, M. T. and O'Doherty, J. V. (2014). Effect of seaweed-derived laminarin and fucoidan and zinc oxide on gut morphology, nutrient transporters, nutrient digestibility, growth performance and selected microbial populations in weaned pigs. *British Journal of Nutrition*. Cambridge University Press 111(9): 1577-1585. DOI: 10.1017/S0007114513004224.

Heim, G., O'Doherty, J. V., O'Shea, C. J., Doyle, D. N., Egan, A. M., Thornton, K. and Sweeney, T. (2015a). Maternal supplementation of seaweed-derived polysaccharides improves intestinal health and immune status of suckling piglets. *Journal of Nutritional Science*. Cambridge University Press 4: e27-e27. DOI: 10.1017/jns.2015.16.

Heim, G., Sweeney, T., O'Shea, C. J., Doyle, D. N. and O'Doherty, J. V. (2015b). Effect of maternal dietary supplementation of laminarin and fucoidan, independently or in combination, on pig growth performance and aspects of intestinal health. *Animal Feed Science and Technology* 204: 28-41. DOI: 10.1016/j.anifeedsci.2015.02.007.

Herman, E. M. and Schmidt, M. A. (2016). The potential for engineering enhanced functional-feed soybeans for sustainable aquaculture feed. *Frontiers in Plant Science* 7: 440. Available at: https://www.frontiersin.org/article/10.3389/fpls.2016.00440.

Holdt, S. L. and Kraan, S. (2011). Bioactive compounds in seaweed: functional food applications and legislation. *Journal of Applied Phycology* 23(3): 543-597. DOI: 10.1007/s10811-010-9632-5.

Honya, M., Mori, H., Anzai, M., Araki, Y. and Nisizawa, K. (1999). Monthly changes in the content of fucans, their constituent sugars and sulphate in cultured *Laminaria japonica*. In: *Proceedings of the Sixteenth International Seaweed Symposium*, pp. 411-416. DOI: 10.1007/978-94-011-4449-0_49.

Jones, R. T., Blunden, G. and Probert, A. J. (1979). Effects of dietary Ascophyllum nodosum on blood parameters of rats and pigs. *Botanica Marina* 22: 393-402.

Jones, C. K., Gabler, N. K., Main, R. G. and Patience, J. F. (2012). Characterizing growth and carcass composition differences in pigs with varying weaning weights and postweaning performance1. *Journal of Animal Science* 90(11): 4072-4080. DOI: 10.2527/jas.2011-4793.

Khan, N., Ryu, K. Y., Choi, J. Y., Nho, E. Y., Habte, G., Choi, H., Kim, M. H., Park, K. S. and Kim, K. S. (2015). Determination of toxic heavy metals and speciation of arsenic in seaweeds from South Korea. *Food Chemistry* 169: 464-470. DOI: 10.1016/j.foodchem.2014.08.020.

Koreleski, J., Świątkiewicz, S. and Arczewska, A. (2010). The effect of dietary potassium and sodium on performance, carcass traits, and nitrogen balance and excreta moisture in broiler chicken. *Journal of Animal and Feed Sciences* 19(2): 244-256. DOI: 10.22358/jafs/66285/2010.

Kraan, S. (2013). 6 - Pigments and minor compounds in algae. In: Domínguez HBT-FI from A for F and N (Ed.) *Functional Ingredients from Algae for Foods and Nutraceuticals. Woodhead Publishing Series in Food Science, Technology and Nutrition*. Woodhead Publishing, pp. 205-251. DOI: 10.1533/9780857098689.1.205.

Lafarga, T., Acién-Fernández, F. G. and Garcia-Vaquero, M. (2020). Bioactive peptides and carbohydrates from seaweed for food applications: natural occurrence, isolation, purification, and identification. *Algal Research* 48. DOI: 10.1016/j.algal.2020.101909: 101909.

Lahaye, M. (1991). Marine algae as sources of fibres: determination of soluble and insoluble dietary fibre contents in some 'sea vegetables'. *Journal of the Science of Food and Agriculture*. John Wiley & Sons, Ltd 54(4): 587–594. DOI: 10.1002/jsfa.2740540410.

Landa-Cansigno, C., Hernández-Carmona, G., Arvizu-Higuera, D. L., Muñoz-Ochoa, M. and Hernández-Guerrero, C. J. (2017). Bimonthly variation in the chemical composition and biological activity of the brown seaweed Eisenia arborea (Laminariales: Ochrophyta) from Bahía Magdalena, Baja California Sur, Mexico. *Journal of Applied Phycology* 29(5): 2605–2615. DOI: 10.1007/s10811-017-1195-2.

Leonard, S. G., Sweeney, T., Bahar, B., Lynch, B. P. and O'Doherty, J. V. (2011). Effects of dietary seaweed extract supplementation in sows and post-weaned pigs on performance, intestinal morphology, intestinal microflora and immune status. *British Journal of Nutrition*. Cambridge University Press 106(5): 688–699. DOI: 10.1017/S0007114511000997.

Lourenço, S. O., Barbarino, E., De-Paula, J. C., Pereira, L. OdS. and Marquez, U. M. L. (2002). Amino acid composition, protein content and calculation of nitrogen-to-protein conversion factors for 19 tropical seaweeds. *Phycological Research*. John Wiley & Sons, Ltd 50(3): 233–241. DOI: 10.1111/j.1440-1835.2002.tb00156.x.

Lozano, I., Wacyk, J. M., Carrasco, J. and Cortez-San Martín, M. A. (2016). Red macroalgae Pyropia columbina and Gracilaria chilensis: sustainable feed additive in the Salmo salar diet and the evaluation of potential antiviral activity against infectious salmon anemia virus. *Journal of Applied Phycology* 28(2): 1343–1351. DOI: 10.1007/s10811-015-0648-8.

Mæhre, H. K., Malde, M. K., Eilertsen, K. E. and Elvevoll, E. O. (2014). Characterization of protein, lipid and mineral contents in common Norwegian seaweeds and evaluation of their potential as food and feed. *Journal of the Science of Food and Agriculture*. John Wiley & Sons, Ltd 94(15): 3281–3290. DOI: 10.1002/jsfa.6681.

Makkar, H. P. S., Tran, G., Heuzé, V., Giger-Reverdin, S., Lessire, M., Lebas, F. and Ankers, P. (2016). Seaweeds for livestock diets: a review. *Animal Feed Science and Technology* 212: 1–17. DOI: 10.1016/j.anifeedsci.2015.09.018.

Manns, D., Nielsen, M. M., Bruhn, A., Saake, B. and Meyer, A. S. (2017). Compositional variations of brown seaweeds Laminaria digitata and Saccharina latissima in Danish waters. *Journal of Applied Phycology* 29(3): 1493–1506. DOI: 10.1007/s10811-017-1056-z.

McAlpine, P., O'Shea, C. J., Varley, P. F., Flynn, B. and O'Doherty, J. V. (2012). The effect of seaweed extract as an alternative to zinc oxide diets on growth performance, nutrient digestibility, and fecal score of weaned piglets. *Journal of Animal Science* 90 (Suppl. 4): 224–226. DOI: 10.2527/jas.53956.

McDonnell, P., Figat, S. and O'Doherty, J. V. (2010). The effect of dietary laminarin and fucoidan in the diet of the weanling piglet on performance, selected faecal microbial populations and volatile fatty acid concentrations. *Animal* Cambridge University Press 4(4): 579–585. DOI: 10.1017/S1751731109991376.

Michalak, I. and Chojnacka, K. (2014). Algal extracts: technology and advances. *Engineering in Life Sciences*. John Wiley & Sons, Ltd 14(6): 581-591. DOI: 10.1002/elsc.201400139.

Michiels, J., Skrivanova, E., Missotten, J., Ovyn, A., Mrazek, J., De Smet, S. and Dierick, N. (2012). Intact brown seaweed (Ascophyllum nodosum) in diets of weaned piglets: effects on performance, gut bacteria and morphology and plasma oxidative status. *Journal of Animal Physiology and Animal Nutrition*. John Wiley & Sons, Ltd 96(6): 1101-1111. DOI: 10.1111/j.1439-0396.2011.01227.x.

Miranda, M., Lopez-Alonso, M. and Garcia-Vaquero, M. (2017). Macroalgae for functional feed development: applications in aquaculture, ruminant and swine feed industries. In: *Seaweeds: Biodiversity, Environmental Chemistry and Ecological Impacts*, pp. 133-153. Available at: https://www.scopus.com/inward/record.uri?eid=2-s2.0-850 48822636&partnerID=40&md5=97e44f36e4ee7081f3c8bb790bae7eba.

Mols-Mortensen, A., Ortind, E. áG., Jacobsen, C. and Holdt, S. L. (2017). Variation in growth, yield and protein concentration in Saccharina latissima (Laminariales, Phaeophyceae) cultivated with different wave and current exposures in the Faroe Islands. *Journal of Applied Phycology* 29(5): 2277-2286. DOI: 10.1007/s10811-017-1169-4.

Morais, T., Inácio, A., Coutinho, T., Ministro, M., Cotas, J., Pereira, L. and Bahcevandziev, K. (2020). Seaweed potential in the animal feed: a review. *Journal of Marine Science and Engineering*. MDPI AG 8(8). DOI: 10.3390/JMSE8080559.

Moroney, N. C., O'Grady, M. N., O'Doherty, J. V. and Kerry, J. P. (2012). Addition of seaweed (Laminaria digitata) extracts containing laminarin and fucoidan to porcine diets: influence on the quality and shelf-life of fresh pork. *Meat Science* 92(4): 423-429. DOI: 10.1016/j.meatsci.2012.05.005.

Moroney, N. C., O'Grady, M. N., Robertson, R. C., Stanton, C., O'Doherty, J. V. and Kerry, J. P. (2015). Influence of level and duration of feeding polysaccharide (laminarin and fucoidan) extracts from brown seaweed (Laminaria digitata) on quality indices of fresh pork. *Meat Science* 99: 132-141. DOI: 10.1016/j.meatsci.2014.08.016.

Nielsen, M. M., Manns, D., D'Este, M., Krause-Jensen, D., Rasmussen, M. B., Larsen, M. M., Alvarado-Morales, M., Angelidaki, I. and Bruhn, A. (2016). Variation in biochemical composition of Saccharina latissima and Laminaria digitata along an estuarine salinity gradient in inner Danish waters. *Algal Research* 13: 235-245. DOI: 10.1016/j.algal.2015.12.003.

O'Doherty, J. V., McDonnell, P. and Figat, S. (2010). The effect of dietary laminarin and fucoidan in the diet of the weanling piglet on performance and selected faecal microbial populations. *Livestock Science* 134(1-3): 208-210. DOI: 10.1016/j.livsci.2010.06.143.

O'Shea, C. J., McAlpine, P., Sweeney, T., Varley, P. F. and O'Doherty, J. V. (2014). Effect of the interaction of seaweed extracts containing laminarin and fucoidan with zinc oxide on the growth performance, digestibility and faecal characteristics of growing piglets. *British Journal of Nutrition*. Cambridge University Press 111(5): 798-807. DOI: 10.1017/S0007114513003280.

Øverland, M., Mydland, L. T. and Skrede, A. (2019). Marine macroalgae as sources of protein and bioactive compounds in feed for monogastric animals. *Journal of the Science of Food and Agriculture*. John Wiley & Sons, Ltd 99(1): 13-24. DOI: 10.1002/jsfa.9143.

Rajauria, G. (2018). Optimization and validation of reverse phase HPLC method for qualitative and quantitative assessment of polyphenols in seaweed. *Journal* of *Pharmaceutical and Biomedical Analysis* 148: 230–237. DOI: 10.1016/j. jpba.2017.10.002.

Rajauria, G., Draper, J., McDonnell, M. and O'Doherty, J. V. (2016). Effect of dietary seaweed extracts, galactooligosaccharide and vitamin E supplementation on meat quality parameters in finisher pigs. *Innovative Food Science and Emerging Technologies* 37: 269–275. DOI: 10.1016/j.ifset.2016.09.007.

Rattigan, R., Sweeney, T., Vigors, S., Thornton, K., Rajauria, G. and O'Doherty, A. J. V. (2019). The effect of increasing inclusion levels of a fucoidan-rich extract derived from Ascophyllum nodosum on growth performance and aspects of intestinal health of pigs post-weaning. *Marine Drugs* 17(12) . DOI: 10.3390/md17120680.

Rattigan, R., Sweeney, T., Maher, S., Thornton, K., Rajauria, G. and O'Doherty, J. V. (2020). Laminarin-rich extract improves growth performance, small intestinal morphology, gene expression of nutrient transporters and the large intestinal microbial composition of piglets during the critical post-weaning period. *British Journal of Nutrition*. Cambridge University Press 123(3): 255–263. DOI: 10.1017/ S0007114519002678.

Regulation EC (2011) Commission Regulation (EU) No 574/2011 of 16 June 2011 amending Annex I to Directive 2002/32/EC of the European Parliament and of the Council as regards maximum levels for nitrite, melamine. Ambrosia spp. and carry-over of certain coccidiostats and histom. *Official Journal of the European Union, L* 159: 7–24.

Rey-Crespo, F., López-Alonso, M. and Miranda, M. (2014). The use of seaweed from the Galician coast as a mineral supplement in organic dairy cattle. *Animal* 8(4): 580–586. DOI: 10.1017/S1751731113002474.

Rioux, L. E., Turgeon, S. L. and Beaulieu, M. (2010). Structural characterization of laminaran and galactofucan extracted from the brown seaweed Saccharina longicruris. *Phytochemistry* 71(13): 1586–1595. DOI: 10.1016/j.phytochem.2010.05.021.

Roleda, M. Y., Marfaing, H., Desnica, N., Jónsdóttir, R., Skjermo, J., Rebours, C. and Nitschke, U. (2019). Variations in polyphenol and heavy metal contents of wild-harvested and cultivated seaweed bulk biomass: health risk assessment and implication for food applications. *Food Control* 95: 121–134. DOI: 10.1016/j.foodcont.2018.07.031.

Ronan, J. M., Stengel, D. B., Raab, A., Feldmann, J., O'Hea, L., Bralatei, E. and McGovern, E. (2017). High proportions of inorganic arsenic in Laminaria digitata but not in Ascophyllum nodosum samples from Ireland. *Chemosphere* 186: 17–23. DOI: 10.1016/j.chemosphere.2017.07.076.

Rose, M., Lewis, J., Langford, N., Baxter, M., Origgi, S., Barber, M., MacBain, H. and Thomas, K. (2007). Arsenic in seaweed—forms, concentration and dietary exposure. *Food and Chemical Toxicology* . Elsevier 45(7): 1263–1267.

Ruiz, Á. R., Gadicke, P., Andrades, S. M. and Cubillos, R. (2018). Supplementing nursery pig feed with seaweed extracts increases final body weight of pigs. *Austral Journal of Veterinary Sciences*. Universidad Austral de Chile. Facultad de Ciencias Veterinarias 50(2): 83–87. DOI: 10.4067/S0719-81322018000200083.

Sauvageau, C. (Camille) (1920). *Utilisation Des Algues Marines*. Paris: O. Doin. Available at: https://www.biodiversitylibrary.org/item/122691.

Schiener, P., Black, K. D., Stanley, M. S. and Green, D. H. (2015). The seasonal variation in the chemical composition of the kelp species Laminaria digitata, Laminaria Hyperborea, Saccharina latissima and Alaria esculenta. *Journal of Applied Phycology* 27(1): 363–373. DOI: 10.1007/s10811-014-0327-1.

Schiener, P., Zhao, S., Theodoridou, K., Carey, M., Mooney-McAuley, K. and Greenwell, C. (2017). The nutritional aspects of biorefined Saccharina latissima, Ascophyllum nodosum and Palmaria palmata. *Biomass Conversion and Biorefinery* 7(2): 221–235. DOI: 10.1007/s13399-016-0227-5.

Silva, D. M., Valente, L. M. P., Sousa-Pinto, I., Pereira, R., Pires, M. A., Seixas, F. and Rema, P. (2015). Evaluation of IMTA-produced seaweeds (Gracilaria, Porphyra, and Ulva) as dietary ingredients in Nile tilapia, Oreochromis niloticus L., juveniles. Effects on growth performance and gut histology. *Journal of Applied Phycology* 27(4): 1671–1680. DOI: 10.1007/s10811-014-0453-9.

Stévant, P., Marfaing, H., Rustad, T., Sandbakken, I., Fleurence, J. and Chapman, A. (2017). Nutritional value of the kelps Alaria esculenta and Saccharina latissima and effects of short-term storage on biomass quality. *Journal of Applied Phycology* 29(5): 2417–2426. DOI: 10.1007/s10811-017-1126-2.

Stiger-Pouvreau, V., Bourgougnon, N. and Deslandes, E. (2016). Carbohydrates from seaweeds. In: Fleurence, J. (Translator) and Levine IBT-S in H and DP (eds) *Seaweed in Health and Disease Prevention*. San Diego: Academic Press, pp. 223–274. DOI: 10.1016/B978-0-12-802772-1.00008-7.

Suetsuna, K. and Nakano, T. (2000). Identification of an antihypertensive peptide from peptic digest of wakame (Undaria pinnatifida). *The Journal of Nutritional Biochemistry* 11(9): 450–454. DOI: 10.1016/S0955-2863(00)00110-8.

Swanson, J. C. (1995). Farm animal well-being and intensive production systems2. *Journal of Animal Science* 73(9): 2744–2751. DOI: 10.2527/1995.7392744x.

Sweeney, T. and O'Doherty, J. V. (2016). Marine macroalgal extracts to maintain gut homeostasis in the weaning piglet. *Domestic Animal Endocrinology* 56 (Suppl.): S84–S89. DOI: 10.1016/j.domaniend.2016.02.002.

Van Boeckel, T. P., Brower, C., Gilbert, M., Grenfell, B. T., Levin, S. A., Robinson, T. P., Teillant, A. and Laxminarayan, R. (2015). Global trends in antimicrobial use in food animals. *Proceedings of the National Academy of Sciences of the United States of America*. National Academy of Sciences 112(18): 5649–5654. DOI: 10.1073/pnas.1503141112.

Vigors, S., O'Doherty, J. V., Rattigan, R., McDonnell, M. J., Rajauria, G. and Sweeney, T. (2020). Effect of a laminarin rich macroalgal extract on the caecal and colonic microbiota in the post-weaned pig. *Marine Drugs* 18(3). DOI: 10.3390/md18030157.

Vigors, S., O'Doherty, J., Rattigan, R. and Sweeney, T. (2021). Effect of supplementing seaweed extracts to pigs until d35 post-weaning on performance and aspects of intestinal health. *Marine Drugs* 19(4) . DOI: 10.3390/md19040183.

Walsh, A. M., Sweeney, T., O'Shea, C. J., Doyle, D. N. and O'Doherty, J. V. (2013a). Effect of dietary laminarin and fucoidan on selected microbiota, intestinal morphology and immune status of the newly weaned pig. *British Journal of Nutrition*. Cambridge University Press 110(9): 1630–1638. DOI: 10.1017/S0007114513000834.

Walsh, A. M., Sweeney, T., O'Shea, C. J., Doyle, D. N. and O'Doherty, J. V. O. (2013b). Effect of supplementing varying inclusion levels of laminarin and fucoidan on growth performance, digestibility of diet components, selected faecal microbial populations and volatile fatty acid concentrations in weaned pigs. *Animal Feed Science and Technology* 183(3–4): 151–159. DOI: 10.1016/j.anifeedsci.2013.04.013.

Wan, J., Jiang, F., Xu, Q., Chen, D. and He, J. (2016). Alginic acid oligosaccharide accelerates weaned pig growth through regulating antioxidant capacity, immunity and intestinal development. *RSC Advances*. The Royal Society of Chemistry 6(90): 87026-87035. DOI: 10.1039/C6RA18135J.

Wan, J., Zhang, J., Chen, D., Yu, B. and He, J. (2017). Effects of alginate oligosaccharide on the growth performance, antioxidant capacity and intestinal digestion-absorption function in weaned pigs. *Animal Feed Science and Technology* 234: 118-127. DOI: 10.1016/j.anifeedsci.2017.09.006.

Wan, J., Zhang, J., Chen, D., Yu, B., Huang, Z., Mao, X., Zheng, P., Yu, J. and He, J. (2018). Alginate oligosaccharide enhances intestinal integrity of weaned pigs through altering intestinal inflammatory responses and antioxidant status. *RSC Advances*. The Royal Society of Chemistry 8(24): 13482-13492. DOI: 10.1039/C8RA01943F.

Wang, T., Jónsdóttir, R., Kristinsson, H. G., Hreggvidsson, G. O., Jónsson, J. Ó., Thorkelsson, G. and Ólafsdóttir, G. (2010). Enzyme-enhanced extraction of antioxidant ingredients from red algae Palmaria palmata. *LWT–Food Science and Technology* 43(9): 1387-1393. DOI: 10.1016/j.lwt.2010.05.010.

Wilkinson, M. (1992). *Seaweed Resources in Europe: Uses and Potential*. Guiry, M. D. and Blunden, G. (Eds). Chichester: John Wiley & Sons Ltd, 1991. xi+432pp. Price: £65.00. ISBN 0 471 92947 6. Aquatic Conservation: Marine and Freshwater Ecosystems 2(2). John Wiley & Sons, Ltd: 209-210. DOI: 10.1002/aqc.3270020206.

Yamada, M., Yamamoto, K., Ushihara, Y. and Kawai, H. (2007). Variation in metal concentrations in the brown alga Undaria pinnatifida in Osaka Bay, Japan. *Phycological Research*. John Wiley & Sons, Ltd 55(3): 222-230. DOI: 10.1111/j.1440-1835.2007.00465.x.

Yuan, Y., Zhang, J., Fan, J., Clark, J., Shen, P., Li, Y. and Zhang, C. (2018). Microwave assisted extraction of phenolic compounds from four economic brown macroalgae species and evaluation of their antioxidant activities and inhibitory effects on α-amylase, α-glucosidase, pancreatic lipase and tyrosinase. *Food Research International* 113: 288-297. DOI: 10.1016/j.foodres.2018.07.021.

Chapter 2

High protein corn fermentation products for swine derived from corn ethanol production

Peter E. V. Williams, FluidQuipTechnologies, USA

1 Introduction

2 Distillers dried grains with solubles

3 Corn fermented protein

4 Challenges in producing corn fermented protein

5 Case study: standardized ileal digestibility of CFP for pigs

6 Case study: concentrations of digestible and metabolizable energy in CFP products fed to pigs

7 Case study: effects of adding phytase on availability of calcium and phosphorus in corn-fermented products

8 Case study: effects on performance of pigs of inclusion of corn-fermented protein in the diet

9 Conclusion

10 References

1 Introduction

Crops are a primary source of protein, and for many years, soybean meal (SBM) has been the protein of choice for livestock feed. However, the pleiotropic relationship between nitrogen and protein content has, over time, resulted in the reduction in protein content of crops such as soybean and corn (maize) as the economics of grain production has prioritized increased yield. This in turn has exacerbated the need for higher concentration protein supplements for feed formulation. In addition, there has also been a movement to reduce the reliance on soy products and replace them with more sustainable sources of protein and particularly vegetable proteins with the absence of anti-nutritional factors.

Decisions on how a feed is formulated are already being driven by the sustainability characteristics of individual feed ingredients. Sustainability (meeting energy and carbon emissions reduction targets) alongside

http://dx.doi.org/10.19103/AS.2024.0140.20

demographics (e.g. the age of those in the labour market) have been identified by the International Feed Industry Federation as key global issues facing the industry. Indexes of sustainability such as green house gas (GHG) and land use change (LUC) metrics are likely to be two new parameters that will become components of feed specification sheets in addition to the nutritional data that has been the norm.

To improve sustainability, processed animal proteins are being reintroduced into feed formulations and work is ongoing to produce a range of single-cell proteins, insect protein and microbial protein. Attention is also being paid to alternative crops high in protein. However, there are considerable challenges to growing new alternative crops on any scale. It takes several years to breed an elite variety to achieve maximum production in different geographic locations with different environmental conditions. Logistical factors such as acquiring appropriate machinery and inputs, as well as acquiring agronomic knowledge to successfully grow a new crop, all serve to favour the status quo. In addition, new crops require land and resources, and potentially compete with existing crop production, contributing to GHG emissions from agriculture as well as indirectly exacerbating climate change from changes in land use (through deforestation). In contrast, plant-based coproducts, secondary products generated during a manufacturing process, have a unique position in that they do not demand additional acreage and do not compete with human food consumption (Mottet et al., 2017).

2   Distillers dried grains with solubles

Corn is the major cereal crop grown in the USA and widely cultivated globally. Corn is traditionally considered an energy crop grown for the energy produced in the form of starch and oil but with a limited amount of protein. As well as being grown as a food crop, corn has also become an important feedstock for the sustainable biofuel industry, with over 100 million tons of corn annually processed into ethanol biofuel in the USA. A by-product of ethanol production is distillers dried grains with solubles (DDGS). This is the dried residue remaining after the starch fraction of corn is fermented with yeasts to produce ethanol. Following fermentation, the ethanol is removed by distillation and the remaining fermentation residues are dried to become DDGS. The growth in ethanol production from corn has resulted in a plentiful supply of DDGS as a co-product which has found a valuable market as a livestock feed. However, it is important to note that DDGS, as a by-product of the dry grind ethanol industry, was never designed as a livestock feed but emerged as a convenient means of marketing the residue from the ethanol production process. This creates a potential opportunity to further develop DDGS as a livestock feed product.

More broadly, the dry grind ethanol industry benefits from several important features as a source of animal feed. A dry grind ethanol plant has established logistics for the delivery and dispatch of grain feedstock and any feed products. Individual dry grind ethanol plants benefit from economies of scale, with large plants processing more than one million tons of grain per annum. Ethanol plants are also integrated bioprocessing facilities typically with integrated heat and power generation. These facilities are prime sites for the installation of processing technology to produce new products with higher nutritional value than traditional DDGS products.

3 Corn fermented protein

There are currently four different commercial processes to produce high concentration corn fermented protein (CFP) products (approximately >50% crude protein (CP)) (Fig. 1). By producing products with higher levels of protein compared with conventional DDGS, these technologies offer the potential to diversify and expand the revenue streams of the bioethanol plant in addition to providing new protein products for the feed industry.

The four processes are an evolution of the wet grind milling process and the separation of protein and fibre to produce separate high-value concentrated streams of protein and fibre. Mechanically separated CFP is produced from whole stillage when portions of fibre and oil are removed, concentrating residual grain proteins and yeast by methods involving centrifugation and washing. One version of CFP processing involves the separation of all fractions

		Pre Fermentation separation	Fermented	Yeast	Approx crude protein %
Empyreal		**Protein concentrate**			**75**
CFP mechanically separated	**SEQUENCE™**				**60**
	GPRE NexPro BP50 A+Pro AltiPro				**>48**
CFP flocculant separated	**ProCap Gold**				**52**
CFP	**ANDVantage ANDVantage PROTOMAX**				**40 50 50**

	full exposure to >50 fermentation
	partial exposure to fermentation
	approx 25% spent yeast in dry matter
	undefined yeast content

Figure 1 Technologies for producing a range of commercial high protein feed products.

post- fermentation with all components of the product exposed to fermentation: the Maximised Stillage Co-products (MSC™) process. A second version of CFP manufacture incorporates pre-fermentation grinding and separation of the whole grain. A third recovers the protein via flocculation and a fourth employs electrostatic forces to separate fibre from protein. Key differences between these processes are identified in Fig. 1.

The MSC™ process separates the protein from the whole stillage by mechanical centrifugation and alternate washing steps (Fig. 2). An important element to grasp is that the dry grind ethanol process starts with a 50-70 h fermentation of ground corn. The ground grain is exposed to moisture, raised temperature and enzymatic hydrolysis to facilitate the conversion of starch into ethanol. There has been little focus on the impact of the fermentation process on residual components of the grain, whether fermentation per se has a nutritional impact on the fibrous components found in whole stillage or the composition of protein components. Other than testing of phytase e.g. as a means of potentially increasing protein recovery, the potential value of fermentation in improving product functionality is unexplored. Xu et al. (2020) listed such potential improvements as improved protein digestibility, increased number of small peptides, increased energy digestibility, reduced fibre levels, improved nutritional value, bioavailability, weight gain and feed conversion rate. Work is in progress to investigate whether additional functional benefits can be obtained for CFP products via the use of fermentation.

In addition, little attention has been given to the fact that, prior to fermentation, there is a significant *in situ* generation of yeast to facilitate the fermentation process and that spent yeast at the end of the ethanol process is

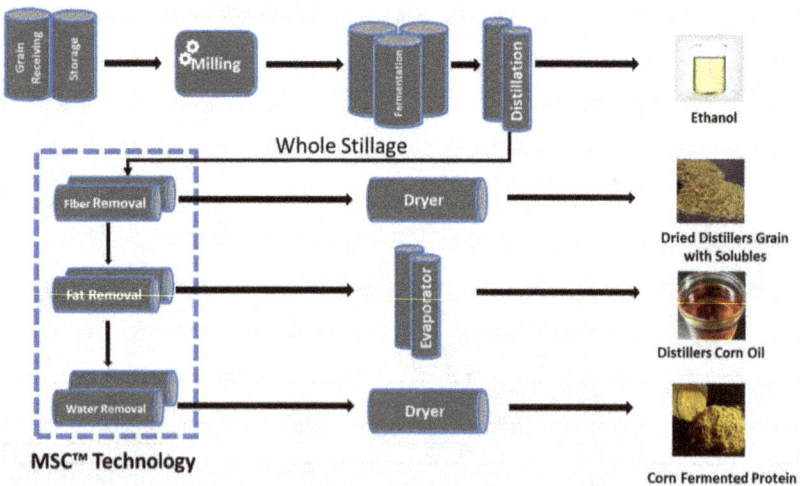

Figure 2 The Maximised Stillage Coproduct Process.

a valuable nutritional component of DDGS. Indeed, the spent yeast of DDGS has been calculated to be approximately 6% of DDGS dry matter (Han and Liu, 2010).

During fermentation, there is a significant net generation of protein from the yeast that is grown in the fermenter to facilitate fermentation. In addition to recovering corn protein, the MSC™ process also recovers a high proportion of spent yeast in particular cell wall components. Whole yeast cells and yeast cell components are recognized as valuable feed nutrient supplements (Shurson, 2018). This produces an exceptionally high-protein CFP product (up to 60% CP; ultra high Pro+, now given the trade name SEQUENCE) that has now been extensively and very successfully tested as a protein supplement for several different species including pigs.

A key challenge has been to measure the quantity of either yeast or yeast components in a dried material such as DDGS. The measurement of the yeast cell wall mannose is accepted by the yeast industry as a proxy measure for yeast cell material. Yeast generated in fermentation accompanies stillage in the ethanol process and passes through the distillation columns where the yeast cells are lysed, releasing the intracellular components including nucleotides. The components of yeast cell walls (mannan oligosaccharides and β-glucans) can be measured in dried CFP products. Using this proxy measure, the yeast cell wall material represents approximately 20–27% of the dry matter content of CFP. Approximately 9% of the protein in a 52% protein product is derived from yeast cell material. However, it is important to recognize that these spent yeast components are not equivalent to whole and viable yeast cells used as feed supplements though the immuno-stimulatory activity of yeast cell components may be valuable functional components of CFP.

4 Challenges in producing corn fermented protein

There are several challenges in producing a commercial CFP product. One potential safety issue is the use of antibiotics in the ethanol process. As a means of optimizing the efficiency of the yeast ethanol conversion step in the dry grind ethanol process, antibiotics have been used during fermentation to eliminate competitive bacterial activity. In the past, Virginiamycin, Erythromycin and Tylosin have been used routinely. Given that the stillage is exposed to high temperature during distillation there is no residual antibiotic activity in the co-products, but antibiotic residues are detectable. Plants producing CFP can use a number of alternative products to antibiotics and, with attention to improved cleaning in place (CIP), are able to operate without the use of antibiotics to control the fermentation process, allowing them to be compliant with No Antibiotics Ever (NAE) programmes for food and feed production.

Another key aspect of product development is its environmental impact. CFP is produced in a highly competitive, sustainable manner with low GHG emission values compared with a range of different protein products (Fig. 3). Using CFP in ration formulation as a partial replacement of SBM has been reported to reduce the GHG value of feed formulation in diets for turkey poults (Burton, 2021).

Key to commercial sustainability is the ability to produce at scale. Up to 500 k metric tons per annum of a specific formulated feed is a norm for commercial feed production. A batch of the formulated feed of up to 100 tons is made, then mixed in 10-12 tons mixer batches. Storage space for feed ingredients is limited. Generally large-scale integrators are limited to three to four feed bins on site, which limits the number of individual ingredients that can be stored. The availability of feed ingredients must merit bin space.

Key aspects of supply chain performance are Resilience, Redundancy and Reliability (the three R's). Changing a feed formulation, particularly for poultry, can be an expensive and time-consuming exercise. Resilience is key because feed producers will not work with a single source of supply, in case of a plant stoppage. It is essential to have one or more sites of manufacture of the product for there to be an alternative source of supply. For the same reason, there needs to be redundancy in volume production to account for increased demand. With resilience in the supply chain, there is an additional need for reliability (product consistency) across different suppliers. Producing a new novel feed protein ingredient is therefore not without significant challenge. CFP products are currently being produced from 17 different ethanol plants in the USA and

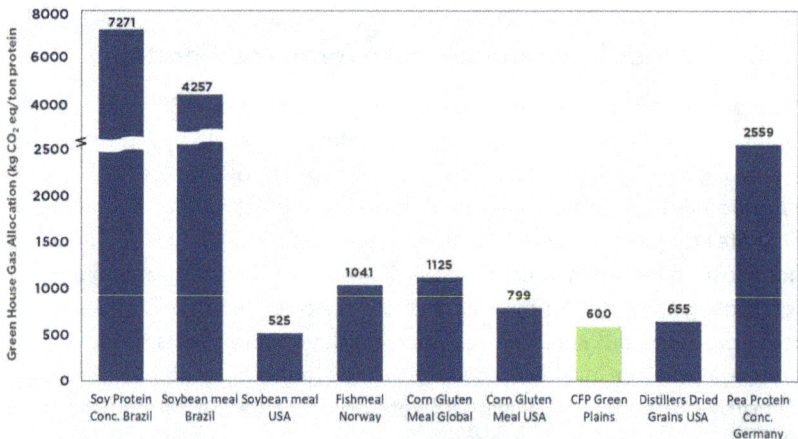

Figure 3 Greenhouse gas allocation values (kg CO_2 eq/ton protein) for a range of different protein products.

Brazil with an annual production capacity close to 1 million tons. Production of the product in Europe is imminent.

Since the first production of CFP 8 years ago standard measures of product composition including amino acid (AA) digestibility have been made using the precision-fed caecectomized rooster assay (PFR) (Parsons, 1985). The assay has been completed on nine separate occasions with samples of CFP obtained during the development of the product, from different ethanol plants as they came online over a period of approximately 6 years to monitor the consistency of the product over time and from different plants. The coefficient of variation of the *in vivo* mean standardized digestibility of the samples of CFP is 2.3%, demonstrating the excellent consistency in the nutritional *in vivo* value of CFP produced from different sites and over a period of approximately 6 years (Table 1).

5   Case study: standardized ileal digestibility of CFP for pigs

An experiment was conducted to compare the standardized ileal digestibility (SID) of CP and AA in high-protein corn fermented products and in the residual DDGSs in comparison with SBM. Nine pigs [initial body weight (BW): 30.56 ± 1.24 kg] were equipped with a T-cannula in the distal ileum (Stein et al., 2007) and were allotted to a 9 × 9 lLatin square design with nine diets and nine periods of seven days. All pigs were limit-fed 3.2 times the maintenance energy requirement (i.e. 197 kcal per kg BW$^{0.60}$; NRC, 2012). Two diets contained bakery meal or SBM as the sole source of AA. Six additional diets contained a mixture of bakery meal-CFP (i.e. at 5%, 10%, 15%, or 20%), bakery meal-ultra high pro+ (i.e. at 10%), or bakery meal-post-MSC DDGS (i.e. at 15%). Values for apparent ileal digestibility (AID), ileal endogenous losses and SID of CP and AA in each diet were calculated (Stein et al., 2007). The basal endogenous losses of CP and AA were calculated from pigs fed with nitrogen-free diet as previously described (Stein et al., 2007).

The AID of Arg, Leu, Thr, Val, Ala and Tyr was greater ($P < 0.05$) in CFP and ultra high pro+ compared with SBM and post-MSC DDGS (Table 2). However, no difference was observed on the SID of CP and AA (except Trp) in SBM, CFP and ultra high pro+. The SID of Lys, Trp, and Asp was less ($P < 0.05$) in Post-MSC DDGS compared with SBM, CFP, and Ultra High Pro+. The SID of His, Val, and Ala also tended to be less ($P < 0.10$) in post-MSC DDGS compared with SBM, CFP and ultra high pro+. Due to increased concentrations of CP and AA in ultra high pro+, concentrations of standardized ileal digestible CP and AA (except Arg, Lys, Trp and Asp) were greater ($P < 0.01$) in ultra high pro+ than in SBM, CFP and post-MSC DDGS (Table 3). The concentration of standardized ileal digestible Arg, Lys, Trp and Asp was greater ($P < 0.01$) in SBM than in

Table 1 Standardized ileal digestibility of amino acids in sample of corn fermented protein

| Amino acid | University Illinois samples 2017–2023 | | | | | | | | | | | | | | | | Univ Georgia | | Mean |
| | CFP1 | | CFP2 | | CFP3 | | CFP4 | | CFP5 | | CFP6 | | CFP7 | | CFP8 | | CFP9 | CFP9 | CFP9 |
	Digest value	SEM³	Digest value	SEM³	Digest value	SEM³	Digest value	SEM³	Digest value	SEM³	Digest value	SEM³	Digest value	SEM³	Digest value	SEM³	Digest value	Digest value	Digest value
Ala	91.5	0.8	92.6	0.6	94.7	0.3	91.8	0.9	91.1	0.8	90	1.6	91.7	0.4	91.6	0.8	89.5	88.6	
Arg	93.3	1	94.6	0.8	98.1	0.5	93.2	0.9	93.6	0.9	91.8	2.3	93.6	0.5	92.3	0.6	91.3	91.5	
Asp	87.7	1.1	87.8	1.3	91.3	0.8	86.1	1.4	84.6	1.3	85.7	1.9	87.2	0.6	87.9	2	85	84.2	
Cys	84.6	2.2	86.3	2.2	91.8	1.1	85.7	1.9	84.5	2.1	78.6	3.7	81.2	1.5	78.9	2.6	79.8	79.5	
Glu	93.1	0.7	93.7	0.6	96.1	0.2	93.1	0.8	92.2	0.7	91	1.7	92.8	0.5	92.9	1.3	91.5	90.8	
Gly																	68.9	52.5	
His	90.2	1.2	90.6	0.8	92.8	0.3	89.2	1.2	87.2	1.1	89.5	1.8	90.4	0.6	91.1	2.2	89.7	90.2	
Ile	90.8	0.8	91.9	0.7	93.2	0.2	90.5	1	89.9	0.9	87.3	2	89.3	0.4	89.3	1	87.6	86	
Leu	94.3	0.5	95.1	0.4	97.1	0.1	94.4	0.6	93.9	0.6	92.9	1.4	94.7	0.3	94	0.4	92.7	92.1	
Lys	83.4	1.5	83.2	1.1	90.5	0.8	81	2.3	80	1.6	85.5	2.4	88.9	0.9	89.2	2.4	80.5	90.7	
Met	91.5	0.4	93.3	0.5	96.9	0.1	92.3	0.9	92	0.6	92.8	1.7	93.9	0.4	93.9	0.7	91.6	90.5	
Phe	92.6	0.8	93.4	0.6	96.1	0.1	92.6	0.8	92.4	0.6	90.6	1.7	92.8	0.3	91.9	0.6	90.6	90.1	
Pro	92.4	0.9	92.4	1.1	95.4	0.2	92	1	90.9	0.7	88.5	1.9	91.1	0.7	89.7	1	90.8	90.2	
Ser	88.5	1.7	89.6	1.9	93	1	89.1	1.3	88.2	1.1	86.5	2.8	89.2	0.9	87.7	1.1	86.4	87.9	
Thr	86.1	1.5	87	1.6	89.9	1.2	85.7	1.6	83.5	1.1	85.5	2.6	87.1	1	87.2	2.3	84.2	84.1	
Trp	94.7	0.9	94.6	0.8	93.6	0.7	92.9	1.2	92.5	0.5	90.2	1.7	90.8	1.1	90.4	1.5	86.3	86.4	
Tyr	93.3	0.9	94.6	0.6	96.9	0.2	93.1	1	92.8	0.8	90.8	1.8	92.1	0.55	91.9	1.1	91.9	91.6	
Val	88.8	1.1	89.6	1.1	95.3	0.3	88.8	1.2	87.6	0.8	89	2	90.8	0.5	89.9	0.7	87.8	86	
Mean	**90.4**		**91.2**		**94.3**		**90.1**		**89.2**		**88.6**		**90.4**		**90**		**87**	**86.3**	**89.7**
															ex GLY		88.1	88.3	cv 2.3

CFP, ultra high pro+, and post-MSC DDGS. Post-MSC DDGS had the least ($P <$ 0.01) concentration of standardized ileal digestible CP and AA among protein sources. The analysed CP and AA values in post-MSC DDGS concur with values reported for conventional DDGS and low-oil DDGS (Fastinger and Mahan, 2006; NRC, 2012; Espinosa et al., 2019). The analysed CP and AA values in CFP and ultra high pro+ are greater compared with values reported for high-protein DDGS (NRC, 2012; Espinosa et al., 2019), which is possibly due to different methods of production used to produce these high-protein corn fermented products. The analysed CP and AA values in CFP is in agreement with values reported by Acosta et al. (2021) for corn protein; however, ultra high pro+ had greater CP and AA compared with values reported by Acosta et al. (2021) for corn protein and values reported by Cristobal et al. (2020) for high- protein DDGS. The SID of AA in SBM was less than values reported by NRC (2012) and Lagos and Stein (2017) for US SBM but agrees with values reported by Espinosa et al. (2020).

The lack of difference observed in SID of CP and AA (except Trp) between SBM and high-protein corn fermented products indicates that these ingredients are of high quality and maybe used as a source of digestible AA in pig diets. The SID of CP and most AA in CFP and ultra high pro+ was not significantly different to that of SBM and was greater compared with values reported for other source of corn protein. The high-protein corn fermented products used in the present experiment were rich sources of CP and AA.

The coefficients of standardized AA digestibility measured in caecetomized cockerels (Table 1) (Parsons et al., 2023) (89-91%) shown for comparative purposes and standardized AA digestibility of AAs for pigs (85-89%) demonstrate that across species the ileal digestibility of these corn-based high protein products is high.

6 Case study: concentrations of digestible and metabolizable energy in CFP products fed to pigs

An experiment was conducted to evaluate the digestible energy (DE) and metabolizable energy (ME) content of high-protein corn fermented products and in post-MSC DDGS. Thirty-two pigs [initial BW: 20.51 ± 1.74 kg] were allotted to four diets with eight replicate pigs per diet using a randomized complete block design. Dietary treatments included a corn-based diet and three diets based on a mixture of corn and each source of high-protein corn fermented product/post-MSC DDGS. Pigs were limit fed at 3.2 times the energy requirement for maintenance (i.e. 197 kcal/kg × $BW^{0.60}$; NRC, 2012). The apparent total tract digestibility (ATTD) of GE was calculated for each diet and the DE and ME in each diet were calculated. The DE and ME in the corn diet were used to calculate the DE and ME in corn. These values were then used

Table 2 Apparent ileal digestibility (AID) (%) of crude protein (CP) and amino acids (AA) in soybean meal (SBM), corn fermented protein product (CFP), Ultra High Pro+, and Post-MSC DDGS1

Item, %	SBM	CFP5*	CFP10*	CFP15*	CFP20*	AV CFP	Ultra high pro+**	Post-MSC DDGS***	SEM	P-value
CP	89.3	83.3	83.0	85.1	84.5	84.0	87.7	74.2	5.3	0.554
Indispensable AA										
Arg	91.1	87.1	89.9	92.2	91.6	90.2	92.3	84.4	3.58	0.551
His	88.7	79.3	81.3	83.7	84.3	82.2	83.9	75.6[z]	2.89	0.099
Ile	88.2	78.2	81.9	82.7	83.7	81.6	82.9	79.5	2.48	0.139
Leu	87.6	88.7	90.5	90.5	90.1	90.0	91.4	86.2	1.89	0.482
Lys	87.2	70.8	76.3	79	78.3	76.1	78.9	48.8[b]	6.37	0.008
Met	88.2	88.9	88.8	88.6	88	88.7	90.4	88	1.93	0.966
Phe	88	84.9	86.2	87.9	88	86.8	88.4	85.2	2.08	0.691
Thr	84	80.8	82.9	82.6	83.1	82.4	83.7	71.9	4.35	0.513
Trp	93.2	75.8	70.9	75.5	78.5	75.2	80.8	58.7[c]	5.46	0.008
Val	86.2	83.9	85.5	84.8	85	84.8	86.1	74.2[y]	3.1	0.094
Ala	85.5	87.8	89.5	88.5	87.7	88.4	90.1	78.0[y]	2.85	0.096
Asp	86.5	76.9	80.2	79.8	81.3	79.6	80.5	67.2[b]	3.59	0.049
Cys	82.4	69.9	74	79.5	80.2	75.9	78	89.6	7.23	0.572
Glu	89.7	88.4	87.5	89	86.9	88.0	90.7	84.7	3	0.837

									SEM	P-value
Gly	90.6	86	86.4	89.6	87.9	87.5	91.8	73.9	9.47	0.888
Ser	88.1	80.4	85.1	86.2	86.7	84.6	88.3	82.6	3.41	0.557
Tyr	88.7	90.3	87.9	90.7	89.7	89.7	91.9	84.6	2.48	0.476
Mean	**87.9**	82.2	83.8	85.3	85.4	**84.2**	86.5	82.8		

[1]Data are least squares means of nine observations per treatment.

a-f Means within a row lacking a common letter are different ($P < 0.05$).

x-z Means within a row lacking a common letter tend to be different ($P < 0.10$).

*Corn fermented protein included at 5%, 10%, 15% and 20% of the diet.

**Ultra high pro+ CFP produced with additional enzymatic processing during fermentation.

***Post MSC DDGS Residual DDGS after MSC process and CFP recovery.

[1]Data are least squares means of nine observations per treatment.

a-d Means within a row lacking a common letter are different ($P < 0.05$).

Table 3 Concentrations of standardized ileal digestible crude protein (CP) and amino acids (AAs) in soybean meal (SBM), corn fermented protein product (CFP), ultra high pro+ and post-MSC DDGS[1] fed to swine

Item g/kg	SBM	CFP5	CFP10	CFP15	CFP20	Ultra high pro	Post-MSC DDGS	SEM	P-value
CP	381.1b	410.6b	407.7b	419.4b	416.4b	501.0a	188.1c	25.54	<0.001
Indispensable AA									
Arg	27.8a	20.5b	21.2b	21.7b	21.6b	21.9b	10.2c	0.82	<0.001
His	9.9c	10.4bc	10.6bc	11.0b	11.0b	11.9a	5.3d	0.36	<0.001
Ile	18.6b	17.5b	18.4b	18.5b	18.8b	20.7a	8.0c	0.51	<0.001
Leu	29.3c	51.8b	52.8b	52.8b	52.6b	68.0a	26.9c	0.95	<0.001
Lys	23.8a	13.7b	14.7b	15.3b	15.2b	14.3b	4.1c	1.1	<0.001
Met	5.1c	9.4b	9.4b	9.4b	9.4b	11.8a	4.3d	0.19	<0.001
Phe	19.7c	22.7b	23.1b	23.5b	23.6b	27.9a	12.2d	0.51	<0.001
Thr	14.1c	15.9bc	16.3ab	16.3ab	16.4ab	17.9a	6.9d	0.73	<0.001
Trp	5.2a	2.8b	3.2b	3.1b	3.1b	5.4a	2.3d	0.2	<0.001
Val	18.3c	22.8b	23.2b	23.1b	23.1b	25.4a	9.3d	0.7	<0.001
Dispensable AA									
Ala	16.1c	31.4b	31.9b	31.6b	31.3b	38.7a	14.3c	0.86	<0.001
Asp	41.5a	28.8b	30.0bc	29.9bc	30.4bc	32.1b	10.6d	1.18	<0.001
Cys	4.5c	6.2bc	6.5b	7.0ab	7.1ab	8.3a	4.6c	0.68	<0.001
Glu	70.4b	71.7b	71.0b	72.2b	70.5b	91.7a	31.2c	2.33	<0.001
Gly	15.7a	17.3a	17.3a	18.0a	17.7a	19.2a	7.8b	1.97	<0.001
Ser	16.3c	17.9bc	19.0b	19.2b	19.3b	22.3a	9.5d	0.72	<0.001
Tyr	14.5c	19.14b	18.6b	19.1b	18.9b	23.2a	8.4d	0.4	<0.001

to calculate the contribution of corn to the corn-high-protein corn fermented product/Post MSC DDGS diets. The DE and ME in high-protein corn fermented products/post-MSC DDGS were calculated by difference (Widmer et al., 2007). Pigs fed the post-MSC DDGS diet had greater ($P < 0.01$) faecal GE loss compared with pigs fed the corn diet or diets containing the high-protein corn fermented products (Table 4). As a result, the ATTD of GE and concentrations of DE and ME in the post-MSC DDGS diet were less ($P < 0.01$) compared with the corn diet, CFP diet or ultra high pro+ diet. Concentrations of DE and ME in diets containing the high-protein corn fermented products were greater ($P < 0.01$) compared with the corn diet.

The ATTD of GE in post-MSC DDGS was less ($P < 0.01$) compared with that of corn, CFP or ultra high pro+. As a result, concentrations of DE and ME (as-fed basis and dry matter basis) in post-MSC DDGS were less ($P < 0.01$) compared with corn, CFP or ultra high pro+. The two high-protein corn fermented products had reduced ($P < 0.01$) ATTD of GE compared with corn; however, due to increased concentration of GE in CFP and ultra high pro+, greater ($P < 0.01$) values for DE and ME (as-fed basis) were observed for CFP and ultra high pro+ compared with corn. Likewise, the DE on a DM basis was also greater in CFP and ultra high pro+ than in corn.

The difference procedure was used to determine concentrations of DE and ME in the two high-protein corn fermented products and post-MSC DDGS, and a consequence of using the difference procedure is that reliable results for test ingredients will be obtained only if the DE and ME of the ingredient included in the corn basal diet are accurate. Values for DE and ME obtained for corn in the present experiment are in close agreement with previous data (NRC, 2012; Espinosa et al., 2019), which gives confidence that calculated values for DE and ME in CFP, ultra high pro+ and post-MSC DDGS are also accurate. Values for DE and ME in post-MSC DDGS used in the experiment were less than published data for conventional DDGS (NRC, 2012), which may be due to increased concentration of insoluble dietary fibre and reduced concentration of fat in post-MSC DDGS compared with conventional DDGS. However, the DE and ME in post-MSC DDGS is in agreement with values for DE and ME in low-oil DDGS (Curry et al., 2016). The observation that CFP and ultra high pro+ had greater concentrations of DE (DM basis) than corn is likely due to increased concentrations of CP and fat as a result of the mechanical separation of solids/protein from the bioethanol whole stillage after the distillation process. Concentrations of DE and ME in post-MSC DDGS were less compared with corn and the two high-protein corn fermented products. Corn had greater ATTD of GE compared with CFP and ultra high pro+; however, DE and ME (as fed basis) and DE on a DM basis were greater in CFP and ultra high pro+ than in corn. The ATTD of GE and concentrations of DE and ME were not different between the two high-protein corn fermented products.

Table 4 Concentration of digestible energy (DE) and metabolizable energy (ME) and apparent total tract digestibility (ATTD) of gross energy in experimental diets and in corn, CFP, ultra high pro+ and post-MSC DDGS[1] fed to swine

Item	Corn	CFP	Ultra high pro+	Post-MSC DDGS	SEM	P-value
Diets						
Gross energy intake, kcal/d	3,728	3,885	3,521	3,493	169	0.268
Gross energy in faeces, kcal/day	378[b]	440[b]	431[b]	535[a]	34	0.007
Gross energy in urine, kcal/day	48	83	71	59	14	0.061
ATTD of gross energy, %	89.9[a]	88.7[ab]	87.9[b]	84.8[c]	0.46	<0.001
DE, kcal/kg	3,289[b]	3,384[a]	3,394[a]	3,205[c]	17	<0.001
ME, kcal/kg	3,242[b]	3,300[a]	3,312[a]	3,142[c]	18	<0.001
Ingredients						
ATTD of gross energy, %	89.9[a]	81.6[b]	80.4[b]	61.8[c]	1.91	<0.001
DE, kcal/kg	3,398[b]	3,987[a]	4,057[a]	2,793[c]	59	<0.001
ME, kcal/kg	3,349[b]	3,707[a]	3,778[a]	2,640[c]	76	<0.001
DE, kcal/kg dry matter	3,914[b]	4,308[a]	4,374[a]	3,201[c]	68	<0.001
ME, kcal/kg dry matter	3,858[a]	4,005[a]	4,072[a]	3,025[b]	86	<0.001

[1]Data are means of eight observations per treatment.
Values within a row lacking a common superscript letter are statistically different ($P < 0.05$).

7 Case study: effects of adding phytase on availability of calcium and phosphorus in corn-fermented products

The 50+ h fermentation which is a critical component of the dry grind bioethanol process is an ideal environment in which to benefit from exogenous enzyme activity to improve nutrient availability in the final product. The standardized total tract digestibility (STTD) of phosphorus (P) in high-protein corn fermented products and in the residual distillers dried grains with solubles (DDGS) produced from a fermentation was evaluated with or without the addition of phytase either into the fermentation medium, or as a dietary supplement (Lopez et al., 2022). Three sources of high-protein corn-fermented products (i.e. CFP, ET-ultra high pro, and ET-ultra high pro+) and two sources of post-MSC DDGS were obtained (Table 5). Two sources of high-protein corn-fermented products (i.e. ET-ultra high pro and ET-ultra high pro+) and one source of post-MSC

Table 5 Analysed nutrient composition of ingredients

Item	CFP	ET[1]-ultra high pro	ET-ultra high pro+	Post-MSC DDGS	ET-post-MSC DDGS
Dry matter, %	91.39	92.55	92.77	87.75	87.26
Gross energy, kcal/kg	4,925	4,885	5,044	4,378	4,520
Ash, %	4.45	3.80	4.28	4.73	3.79
Ca, %	0.04	0.01	0.02	0.03	0.02
Total P, %	0.87	0.57	0.58	1.10	0.70
Phytic acid, %	2.14	<0.14	0.30	2.20	<0.14
Phytate-bound P,[2] %	0.60	<0.04	0.08	0.62	<0.04
Nonphytate P,[3] %	0.27	0.53	0.50	0.48	0.68
Myo-inositol, mg/100 g	29.70	65.50	58.60	77.40	150.00

[1]ET, enzyme-treated.
[2]Calculated as 28.2% of phytic acid (Tran and Sauvant, 2004).
[3]Calculated as total P – phytate-bound P.

DDGS (i.e. ET-post-MSC DDGS) were treated with microbial phytase during the fermentation process, whereas phytase treatment was not included for the other two corn co-products. Five diets were formulated by mixing each source of corn co-product with sucrose and cornstarch, and corn co-product was the only source of P in these diets. Five additional diets were also formulated similar to the previous two diets with the exception that 500 units of phytase per kilogram were added directly to the diets. A standard vitamin and minerals premix was included to meet the nutrient requirements for growing pigs (NRC, 2012).

Eighty pigs [initial BW: 14.38 ± 1.50 kg] were allotted to ten diets with eight replicate pigs per diet. All diets were fed in meal form. Pigs were limit fed at 3.2 times the energy requirement for maintenance (i.e. 197 kcal/kg × $^{BW0.60}$; NRC, 2012). The initial 5 days were considered the adaptation period to the diet, and faecal materials were collected from the feed provided during the following 4 days according to standard procedures using the marker to marker approach (Adeola, 2001).

The apparent total tract digestibility (ATTD) of Ca and P in each diet was calculated (Table 6). By correcting ATTD values for the basal endogenous loss of P (i.e. 190 mg per kg dry matter intake), values for the STTD of P in each diet were also calculated. Due to increased concentration of total P, pigs fed the diet containing CFP and post-MSC DDGS had greater ($P < 0.01$) P intake compared with that of pigs fed diets with other corn co-products. Inclusion of 500 FTU per kilogram of phytase in CFP and post-MSC DDGS diets reduced the

Table 6 Effects of microbial phytase on P balance, apparent total tract digestibility (ATTD), and standardized total tract digestibility (STTD) of P in three sources of high-protein corn fermented products and two sources of post-MSC DDGS fed to growing pigs[1]

Item	ADFI, g/day	P intake, g/day	P in faeces, %	P output, g/day	P absorption, g/day	ATTD of P, %	Basal EPL[2], mg/day	STTD of P[3], %
0 FTU[4]/kg								
CFP	525	1.92	2.13a	0.74b	1.19	62.07d	92.20	66.86d
ET-Ultra High Pro	588	1.49	0.68d	0.30c	1.18	79.74ab	102.52	86.63a
ET-Ultra High Pro+	499	1.33	0.58d	0.23c	1.10	82.95a	87.27	89.53a
Post-MSC DDGS	593	2.62	1.41b	1.01a	1.61	62.23d	101.52	66.10d
ET-Post-MSC DDGS	541	1.68	0.36e	0.34c	1.34	79.46ab	92.42	84.96ab
500 FTU/kg								
CFP	513	1.88	1.51b	0.57b	1.31	70.52c	90.00	75.30c
ET-Ultra High Pro	605	1.53	0.61d	0.27c	1.26	82.29a	105.50	89.19a
ET-Ultra High Pro+	524	1.38	0.58d	0.24c	1.15	82.74a	91.64	89.32a
Post-MSC DDGS	558	2.47	0.89c	0.61b	1.86	75.42bc	95.96	79.31bc
ET-Post-MSC DDGS	523	1.62	0.38e	0.30c	1.31	80.93ab	90.32	86.49a
SEM	34	0.12	0.06	0.06	0.09	2.28	5.86	2.28
P-values								
Corn co-product	0.006	<0.001	<0.001	<0.001	<0.001	<0.001	0.008	<0.001
Phytase	0.788	0.609	<0.001	0.003	0.028	0.001	0.862	0.001
CFP × Phytase	0.777	0.818	<0.001	0.012	0.309	0.024	0.802	0.025

[1]Data are least squares means of seven to eight observations per treatment.
[2]EPL = endogenous P loss. This value was estimated to be at 190 mg/kg DMI (dry matter intake). The daily basal EPL (mg/day) for each diet was calculated by multiplying the EPL (mg/kg DMI) by the daily DMI of each diet (Almeida and Stein, 2010).
[3]Values for STTD calculated by correcting ATTD values for the basal endogenous loss of P (NRC, 2012).
[4]FTU = phytase units.
a-eMeans within a row lacking a common letter are different (P < 0.05).

concentration of P in faeces; however, phytase did not affect the concentration of P in faeces by pigs fed diets containing the ET corn co-products (interaction, $P < 0.01$). Due to increased P intake, P absorption was greater ($P < 0.01$) in pigs fed the diet containing CFP and post-MSC DDGS compared with that of pigs fed diets with other corn co-products regardless of phytase supplementation. Absorption of P also increased ($P < 0.05$) when diets were supplemented with phytase. The ATTD and STTD of P in the CFP and post-MSC DDGS diets increased upon phytase supplementation; however, phytase supplementation did not influence the ATTD and STTD of P in diets containing the enzyme-treated corn co-products (interaction, $P < 0.05$).

Dietary supplementation of phytase increased the ATTD and STTD of P in CFP and Post-MSC DDGS indicates that the exogenous phytase increased P absorption in the pigs (Adeola, 2001). The observed increase in Ca digestibility in experimental diets indicates that phytase was also effective in hydrolyzing the Ca–phytate complexes in the gastrointestinal tract of pigs (Selle et al., 2009). The observation that the ATTD and STTD of P in enzyme-treated corn co-products were greater compared with CFP and Post-MSC DDGS (without enzyme treatment) indicates that the use of phytase during the fermentation process was able to liberate some of the P in the phytic acid molecule. Indeed, this agreed with the observed reduction in the analysed concentration of phytate-bound P and a subsequent increase in the analysed concentration of myo-inositol in corn co-products that were treated with phytase during the fermentation process. The observation that dietary supplementation of phytase did not influence P digestibility in enzyme-treated corn co-products indicates that phytase treatment during the fermentation process is sufficient to maximize P utilization in high-protein corn fermented products and Post-MSC DDGS. Concentration of phytate-bound P was less and myo-inositol concentration was greater in enzyme-treated corn co-products compared with corn co-products without enzyme treatment. As a result, the ATTD and STTD of P in enzyme-treated corn co-products were greater compared with CFP and Post-MSC DDGS (without enzyme treatment). Dietary supplementation of phytase increased the ATTD of Ca in experimental diets and increased the ATTD and STTD of P in CFP and post-MSC DDGS. Addition of phytase to diets did not influence P digestibility in enzyme-treated corn co-products, which likely indicates that phytase treatment during the fermentation process is sufficient to maximize P utilization in corn co-products.

8 Case study: effects on performance of pigs of inclusion of corn-fermented protein in the diet

Formulations of creep feeds for nursery pigs are recognized to contain a wide variety of feed ingredients to stimulate intake and minimize the occurrence of diarrhoea in neonatal pigs. Highly digestible benign proteins, such as spray-dried animal plasma (SDAP), plant proteins such as enzyme-treated SBM or mixtures of high value protein plus additions of components of yeast, are common additions to nursery pig diets. Nutritionists are keen to minimize the number of products in nursery diets (phase 1 diets) that contain recognized anti-nutritional factors such as SBM due to the antigenic effects limiting early feed intake. One of the most appropriate uses of CFP co-products because of its high ME and digestible AA content and lack of anti-nutritional factors is in phase 1 and 2 diets for weaned pigs.

Trial 1: Martindale et al. (2018) determined the effects of increasing dietary levels (0%, 8%, 16%, and 24%) of CFP in nursery, phase 1 (0–14 days post-weaning) and phase 2 (14–28 days post-weaning), with a common corn-SBM diet fed during phase 3 (days 28–35 post-weaning) on growth performance of pigs weaned up to 21 days of age. No differences in average daily gain (ADG), average daily feed intake (ADFI) and gain:feed (G:F) were observed among dietary treatments during phase 1, but pigs fed the 24% CFP diet during phase 2 had reduced ADG and ADFI compared with pigs fed the control (0% CFP) diet. There were no effects on growth performance during phase 3 and the overall 35-day trial. The researchers concluded that adding up to 16% CFP to phase 1 and phase 2 nursery diets does not negatively affect growth performance.

Trial 2: Acosta et al. (2021) evaluated the impact on growth performance and faecal scores when feeding diets containing various amounts of CFP to partially replace SDAP and enzyme-treated SBM (ES) for 21 days post- weaning during phase 1 (days 1–7) and phase 2 (days 8–21). Feeding the control phase 1 diet containing 5% ES and 2.5% SDAP for the first 7 days after weaning tended to increase ADG (0.12 kg/d) and G:F (0.78) compared with feeding diets containing 4.5% ES + 5% CFP and 10% CFP (0.08 kg/days; 0.57 respectively), but there was no significant difference in ADFI among dietary treatments. However in phase 2, the amount of CFP (0–10%) fed from day 8–21 post-weaning had no effect on ADG, ADFI and G:F. All pigs were fed a common corn–SBM diet during phase 3 (day 22–35 post-weaning) and there was no differences in subsequent growth. The results indicate that if 2.5% SDAP is included in the diet adding 5% CFP to phase 1 diets for weaned pigs can partially replace spray-dried porcine plasma providing equivalent performance but this was not the case with ES. Adding 10% CFP to phase 1 nursery diets appeared to be excessive to support optimal growth performance without addition of SDAP or ES. However, adding 10% CFP to phase 2 diets supported acceptable growth performance comparable to feeding the control diet containing 7.5% ES.

Trial 3: The effect of CFP at 7.5% and 15% inclusion rate (replacing soy- derived ingredients on an equivalent CP basis) in piglet weaner diets, formulated to meet current nutrient requirements, was evaluated on piglet growth performance when fed a single diet for 28 days post-weaning. Three treatments included a high-specification control diet containing whey (15%)

and soy (SBM (10%) and Alphasoy 530 (5%)) as the primary protein sources, and two similar diets but with CFP added at 7.5% or 15% at the expense of the soya-derived protein ingredients on an equivalent CP basis. All diets were formulated to be isoenergetic and equivalent in terms of nutrient composition. A total of 240 mixed sex pigs ([Large White × Landrace] × Danish Duroc) were weaned onto the experimental diets at an initial weight of 8.24 ± 1.19 kg (mean ± SD) at 26.4 ± 0.56 days of age. Pig health, faecal consistency and cleanliness scores were recorded daily. Pig performance was in line with expectations for the facility, with growth rates averaging 376 ($\pm$ 56.4) g/pig/day over the course of the study. Health, faecal and cleanliness scores were uniformly good throughout the study, with no differences observed between treatments and there were no pig mortalities during the experiment.

There were no differences in pig growth rates or feed conversion ratios between the three treatments during the first 13 days of the study although there was a tendency for a reduction in ADFI in response to feeding CFP at both levels. After 28 days on trial pigs receiving the diets containing CFP were 1.27 and 2.01 kg lighter than control-fed pigs, respectively ($P < 0.01$). This difference in BW was attributed to a reduction in feed intake in response to CFP inclusion over the course of the study, with those fed CFP 7.5% consuming 11% less feed than the control pigs, and those fed CFP 15% consuming 16% less feed. However there was no difference in pig FCR between the three treatments throughout the trial, indicating that CFP is utilized as efficiently as AlphaSoy 530 and SBM. There was no effect of replacing the soy derived proteins with CFP on piglet health or feed conversion efficiency after 28 days of feeding, indicating that this yeast/grain-based protein ingredient can be utilized as efficiently as a conventional soy derived product. The results indicate that CFP can be utilized as an alternative protein supplement in piglet diets up to 15% inclusion for the first 13 days post-weaning, without compromising piglet health or growth performance. However, upon extending the feeding duration of CFP to 28 days post-weaning, depressions in piglet growth rates were observed at both the 7.5% and 15% inclusion levels. This effect was attributed to reductions in piglet feed intake occurring primarily throughout the final two weeks of the study, which suggest that there was an effect of CFP on diet palatability. The results suggest that in young pigs there is a component of CFP that has a negative impact on feed intake. It is reported that excess Leucine competes with tryptophan (Trp) transport through the blood into the brain which reduces serotonin synthesis and consequently reduces ADFI (Kwon et al., 2019; Yang et al., 2019). It was postulated that an imbalance of AAs caused by the inclusion of corn protein may be responsible for the effect on palatability.

Trial 4: The earlier experiments demonstrated that whilst CFP can be used to replace a range of different protein sources in weaner diets for pigs, there is a limit to CFP inclusion above which there is appetite depression. Diets with a high inclusion levels of high-protein distillers dried grain with solubles (HP-DDGS) or CFP may contain an excess of dietary Leucine (Leu), and this may be responsible for the negative effects on feed intake and reduced growth performance. It was hypothesized that the negative effect of using CFP in diets for weaning pigs could be avoided if the AA imbalance could be corrected by addition of crystalline sources of Valine (Val), Tryptophan (Try) and (or) Isoleucine (Ile) (Stein personal communication). A total of 320 weanling pigs (BW: 6.11 ± 0.66 kg) were randomly allotted to one of 10 dietary treatments in a completely randomized design. A two-phase feeding programme was used, days 1 to 14 (phase 1) and days 15 to 28 (phase 2). Three basal diets were formulated for each phase, a corn-SBM diet and 2 test diets based on corn plus 10% CFP or corn plus 20% CFP. Seven additional diets were formulated by adding Val, Ile, Trp, Val and Ile, Val and Trp, Ile and Trp, or Val, Ile, and Trp to the basal diet with 20% CFP. Phase 1 diets were fed from day 1 to 14 post-weaning and phase 2 diets were from day 15 to 28. ADFI, ADG and average G:F were calculated for each phase and for the overall experiment. Faecal scores were recorded every other day. There was no effect of the treatments on the health of the piglets. Results indicated that inclusion of 10% or 20% CFP in diets reduced ($P < 0.05$) final BW on day 28, ADG and ADFI in phase 2 and for the entire experimental period and G:F for the entire experiment similar to results obtained in trial 3. However, pigs fed SBM and CFP supplemented with the 3 AAs had a greater (+1.72 kg live weight; $P < 0.05$) final BW and ADG compared with pigs fed the other diets. On day 28, pigs fed the diet with 20% CFP and only Val, Val and Trp, or Val, Trp, and Ile had reduced ($P < 0.01$) blood urea N compared with pigs fed the corn-SBM diet or the other CFP-based diets. The fact that high concentrations of leucine negatively affects the utilization and metabolism of valine and isoleucine (Ile), causing a nutritional challenge is recognized (Harris et al., 2004; Cemin et al., 2019a,b; Kwon et al., 2019; Yang et al., 2019) and CFP contains high levels of leucine. In addition, excess Leu competes with tryptophan (Trp) transport through the blood into the brain which reduces serotonin synthesis and consequently reduces ADFI (Kwon et al., 2019; Yang et al., 2019). These results indicate that CFP can be used as an alternative vegetable protein source and included in phase 1 and phase 2 nursery diets at levels up to 16% as long as attention is paid to balancing the AA content of the formulation with additional Val, Trp and Ile.

9 Conclusion

There are currently 17 plants producing CFP in the USA, 2 in South America and 1 in construction in Europe with a combined production capacity of well over 1 million tons of CFP per annum, providing resilience and redundancy in the supply chain. Although the dry grind ethanol process is a global source of medium- quality protein, the opportunity to turn corn into a multi-purpose energy and protein crop is a major breakthrough. The trials carried out to evaluate the use of CFP as an alternative protein supplement in diets for broilers and turkey poults demonstrate that CFP can be used in formulation as an alternative protein.

In order to complete these trials the standardized digestibility and true metabolizable energy were measured in the target species. In all the early growth experiments, the maximum inclusion rate of CFP was restricted to not exceed 10% of dietary inclusion. Since these experiments were exploratory experiments to test the use of the product in diets for poultry and CFP contains approximately 20-27% yeast cell wall material, this was a precautionary measure to not exceed the yeast cell wall dietary inclusion level of 5%. CFP can be used as a dietary replacement for SBM and with appropriate balancing of essential AAs will achieve performance at a minimum equal to the performance of the control diets. However, there is evidence to suggest that either the content of spent yeast material or the combination of spent yeast with a fermented carbohydrate fraction may be beneficial in terms of gut health.

A trial with turkey poults is particularly interesting (Williams, 2024). Overall growth performance of poults across all treatments aligned well with industry standards (Aviagen, 2014), indicating that the diets conformed well to industry standards. However, in terms of improved nitrogen end energy retention, there is evidence to suggest that there may be additional benefits to the partial replacement of SBM with CFP.

CFP contains a functional component of approximately 24% (on a DM basis) spent yeast content. Several nutraceutical components of yeast are reported to have beneficial effects in poultry and may indirectly improve growth performance. β-1,3-glucans, the major component of yeast cell walls, have prebiotic effects due to their ability to bind toxins and pathogens (Vetvicka and Oliveira, 2014). Yeasts contain mannan oligosaccharides which have an overall positive effect on animal performance (Spring et al., 2015) via improving intestinal architecture, physical gut tissue turnover, a change in microbiota or reduction in immune stimulation (Baurhoo et al., 2007). The effects on ceacal VFA concentrations reported in a broiler trial are evidence that the fermented fibre component of CFP has the potential to increase hind gut fermentation although to date there is no evidence to indicate an improvement in the histology of the gut.

Identifying new sources of high-quality vegetable protein that are available for use in animal feed is a priority. High quality vegetable proteins are extensively used for food and can therefore demand a price premium. Identifying a source of high concentration vegetable protein such as CFP that is not available for use in food since it is derived from a non-food industrial process is an opportunity for the feed industry to source an alternative cost competitive product. While SBM is recognized as an excellent protein source, residual trypsin inhibitor, lipoxygenases and antigenic proteins retain a degree of activity post-processing which slightly but significantly reduce dietary protein digestion (Clarke and Wiseman, 2005). Thus, some of the performance improvement in the current study could be related to the reduction in anti-nutritional effects associated with partial replacement of SBM with CFP. However, the results demonstrate that CFP can be used as a dietary replacement for SBM and with appropriate balancing of essential AAs will achieve performance at a minimum equal to the performance of the control diets.

The current programme covering the use of CFP in pigs is part of an extensive programme on the use of the protein in diets for poultry, aquaculture, companion animals and ruminants in addition to poultry. The results mainly reflect the beneficial effects reported in poultry. The development of CFP is an excellent example of how the introduction of tested processing technology with existing commodity feed materials can significantly improve the value a low-quality feed material and significantly increase the flexibility and value of feed materials used by animal nutritionists in feed formulation. The composition of CFP with a high protein content and potential functional characteristics based on the content of spent yeast and fermented fibre, point towards the product being an ideal vegetable protein supplement for swine and in particular for use in the nursery phase 0–21 days of age. There have been few if any new and significant developments in the vegetable protein space for animal feed in the past 20 years. CFP is the first major development in terms of a new commercially viable high protein feed for all classes of livestock and companion animals.

10   References

Acosta, J. P., Espinosa, C. D., Jaworski, N. W. and Stein, H. H. 2021. Corn protein has greater concentrations of digestible amino acids and energy than low-oil corn distillers dried grains with solubles when fed to pigs nut does not affect performance of weanling pigs. *J. Anim. Sci.* 99(7):1–12. doi: 10.1093/jas/skab175.

Adeola, O. 2001. Digestion and balance techniques in pigs. In Lewis, A. J. and Southern, L. L. (eds). *Swine Nutrition*, Second edition, CRC Press, Washington DC, USA.

Aviagen 2014. Ross 308 broiler: nutrition specifications. Available at: http://en.aviagen.com/assets/Tech_Center/Ross_Broiler/Ross308BroilerNutritionSpecs2014-EN.pdf, accessed 21 December 2016.

Baurhoo, B., Phillip, L. and Ruiz-Feria, C. A. 2007. Effects of purified lignin and mannan oligosaccharides on intestinal integrity and microbial populations in the ceca and litter of broiler chickens. *Poult. Sci.* 86(6):1070–1078 doi: 10.1093/ps/86.6.1070.

Burton, E. 2021. Use of an ethanol bio-refinery product as a soy bean alternative in diets for fast-growing meat production species. *Sustainability* 13(19):11019.

Cemin, H. S., Tokach, M. D., Dritz, S. S., Woodworth, J. C., DeRouchey, J. M. and Goodband, R. D. 2019a. Meta-regression analysis to predict the influence of branched-chain and large neutral amino acids on growth performance of pigs. *J. Anim. Sci.* 97(6):2505–2514. doi: 10.1093/jas/skz118.

Cemin, H. S., Tokach, M. D., Woodworth, J. C., Dritz, S. S., DeRouchey, J. M. and Goodband, R. D. 2019b. Branched-chain amino acid interactions in growing pig diets. *Transl. Anim. Sci.* 3(4):1246-1253. doi: 10.1093/tas/txz087.

Clarke, E. and Wiseman, J. 2005. Effects of variability on trypsin inhibitor content of soy bean meals on true and apparent ileal digestibility of amino acids and pancreas size in broiler chicks. *Anim. Feed Sci. Technol.* doi:10.1016/j.anifeedsci.2005.02.012.

Cristobal, M., Acosta, J. P., Lee, S. A. and Stein H. H. 2020. A new source of high-protein distillers dried grains with solubles (DDGS) has greater digestibility of amino acids and energy, but less digestibility of phosphorus, than de-oiled DDGS when fed to growing pigs. *J. Anim. Sci.* 98(7): 1-9.

Curry, S. M., Rojas, O. J. and Stein, H. H. 2016. Concentration of digestible and metabolizable energy and digestibility of energy and nutrients by growing pigs in distillers dried grains with solubles produced in and around Illinois. *The Professional Animal Scientist* 32(5):687-694.

Espinosa, C. D., Fry, R. S., Kocher, M. E. and Stein, H. H. 2019. Effects of copper hydroxychloride and distillers dried grains with solubles on intestinal microbial concentration and apparent ileal and total tract digestibility of energy and nutrients by growing pigs. *J. Anim. Sci.* 97(12):4904-4911. doi: 10.1093/jas/skz340.

Espinosa, C. D., Oliveira, M. S. F., Lagos, L. V., Weeden, T. L., Mercado, A. J. and Stein, H. H. 2020. Nutritional value of a new source of fermented soybean meal fed to growing pigs. *J. Anim. Sci.* 98(12):1-9.

Fastinger, N. D. and Mahan, D. C. 2006. Determination of the ileal amino acid and energy digestibilities of corn distillers dried grains with solubles using grower-finisher pigs. *J. Anim. Sci.* 84(7):1722-1728. doi: 10.2527/jas.2005-308.

Harris, R. A., Joshi, M. and Jeoung, N. H. 2004. Mechanism responsible for regulation of branched-chain amino acid catabolism. *Biochem. Res. Commun.* 313:391-396. doi: 10/1093/jas/skz259.

Han, J. and Liu, K. 2010. Changes in composition and amino acid profile during dry grind ethanol processing from corn and estimation of yeast contribution towards DDGS proteins. *J. Agric. Food Chem.* 58(6):3430-3437.

Kwon, W. B., Touchette, K. J., Simongiovanni, A., Syriopoulos, K., Wessels, A. and Stein, H. H. 2019. Excess dietary leucine in diets for growing pigs reduces growth performance, biological value of protein, protein retention, and serotonin synthesis. *J. Anim. Sci.* 97(10):4282-4292. doi: 10.1093/jas/skz259.

Lagos, L. V. and Stein, H. H. 2017. Chemical composition and amino acid digestibility of soybean meal produced in the United States, China, Argentina, Brazil or India. *J. Anim. Sci.* 95:1626-1636.

Lopez, D. A., Lee, S. A. and Stein, H. H. 2022. Effects of microbial phytase on standardized total tract digestibility of phosphorus in feed phosphates fed to growing pigs. *J. Anim. Sci.* 100(12):skac350.

Martindale, A., Trenhaile-Grannemann, M., Barnett, S., Miller, P. and Burkey, T. 2018. Growth performance of weaned pigs fed high-protein corn product. *J. Anim. Sci.* 96(Suppl. 3):295.

Mottet, A., de Haan, C., Falcuucci, A., Tempio, G., Opio, C. and Gerber, P. 2017. Livestock: on our plates or eating at our table? A new analysis of the feed/food debate. *Glob. Food Sec.* 14:1-8. doi: 10.1016/j.gfs.2017.01.001.

NRC 2012. *Nutrient Requirements of Swine* (11th rev. edn). Washington (DC): The National Academies Press.

Parsons, B. W., Utterback, P. L., Parsons, C. M. and Emmert, J. L. 2023. Standardized amino acid digestibility and true metabolizable energy for several increased protein ethanol co-products using back-end fractionation systems. *Poult. Sci.* 102(2):102329.

Parsons, C. M. 1985. Influence of caecectomy on digestibility of amino acids by roosters fed distillers dried grains with solubles. *J. Agric. Sci.* 104(2):469–472.

Rho, Y., Zhu, C., Kiarie, E. and de Lange, C. F. M. 2017. Standardized ileal digestible amino acids and digestible energy contents in high-protein distiller's dried grains with solubles fed to growing pigs. *J. Anim. Sci.* 95(8):3591–3597. doi: 10.2527/jas.2017.1553.

Scholey, D., Alkhtib, A., Williams, P. and Burton, E. 2023. Corn fermented protein, an alternative protein to soybean meal to support growth in young turkey poults. *J. Appl. Anim. Nutr.* doi: 10.3920/JAAN2023.0002.

Selle, P. H., Cowieson, A. J. and Ravindran, V. 2009. Consequences of calcium interactions with phytate and phytase for poultry and pigs. *Livest. Sci.* 124:126–141.

Shurson, G. C. 2018. Yeast and yeast derivatives in feed additives and ingredients: sources, characteristics, animal responses and quantification methods. *Anim. Feed Sci. Technol.* 235:60–76 . doi: 10.1016/j.anifeedsci.2017.11.010.

Spring, P., Wenk, C., Connolly, A. and Kiers, A. 2015. A review of 733 published trials on Bio-Moss, a mannan oligosaccharide and Actigen, a second generation mannose rich fraction, on farm and companion animals. *J. Appl. Anim. Nutr.* 3:1–11.

Stein, H. H., Sève, B., Fuller, M. F., Moughan, P. J. and de Lange, C. F. M. 2007. Invited review: amino acid bioavailability and digestibility in pig feed ingredients: terminology and application. *J. Anim. Sci.* 85(1):172–180. doi: 10.2527/jas.2005-742.

Vetvicka, V. and Oliveira, C. 2014. β(1-3)(1-6)-D-glucans modulate immune status in pigs: potential importance for efficiency of commercial farming. *Ann. Transl. Med.* 2(2):16. doi: 10.3978/j.issn.2305-5839.2014.01.04.

Widmer, M. R., McGinnis, L. M. and Stein, H. H. 2007. Energy, phosphorus, and amino acid digestibility of high-protein distillers dried grains and corn germ fed to growing pigs. *J. Anim. Sci.* 85(11):2994–3003. doi: 10.2527/jas.2006-840.

Williams, P. E. V. 2024. High protein corn fermentation products for poultry derived from corn ethanol production. In Applegate, T. (ed.). *Advances in Poultry Nutrition*, Burleigh Dodds Science Publishing, Cambridge, UK.

Xu, B., Li, Z., Wang, C., Fu, J., Zhang, Y., Wang, Y. and Lu, Z. 2020. Effects of fermented feed supplementation on pig growth performance: a meta-analysis. Anim. Feed Sci. Technol. 259:114315. doi: 10.1016/j.anifeedsci.2019.114315.

Yang, Z., Urriola, P. E., Hilbrands, A. M., Johnston, L. J. and Shurson, G. C. 2019. Growth performance of nursery pigs fed diets containing increasing levels of a novel high-protein corn distillers dried grains with solubles. *Transl. Anim. Sci.* 3(1):350–358. doi: 10.1093/tas/txy101.

Chapter 3

Developing alternative sources of protein in pig nutrition: insects

Kristy DiGiacomo, The University of Melbourne, Australia

1 Introduction

2 Why insects

3 Current state of insect production

4 Potential insects and current production volumes

5 Production and health responses in pigs

6 Competition for resources

7 Rearing insects on manure

8 Current research

9 Future trends in research

10 Barriers to uptake/challenges to production

11 Conclusion

12 Where to look for further information

13 References

1 Introduction

There is an increased global interest in the production of insects for animal feed and human food. This interest is driven by a desire to reduce environmental footprints and increase sustainability, reduce reliance on traditional proteins (such as soybeans) and imported feed ingredients, reduce feed costs, utilize waste, and create alternative revenue streams (or a combination of each of these). Global pork production is an optimized, efficient and integrated system that utilizes breeding, nutrition, housing, and management to optimize production. An increase in the production of pork and pork products is occurring globally as human populations increase and diets shift to include larger portions of meat, particularly in developing countries. The predicted global pig meat production for 2030 is approx. 45 000 tonnes, which is a sustained increase from the approx. 40 000 tonnes produced in 2010 (OECD and Food

http://dx.doi.org/10.19103/AS.2024.0140.17

Agriculture Organization of the United Nations, 2022). Given that poultry meat production is predicted to increase by approximately 14 000 tonnes in the same period, there is increasing competition for animal feed ingredients (OECD and Food Agriculture Organization of the United Nations, 2022). The recent global coronavirus disease 2019 (COVID-19) pandemic highlighted that pork production systems are vulnerable to external factors, particularly global supply chains, further increasing the need to diversify diet ingredients.

Providing protein in the production of animal feed is a major contributor to the environmental impact of pork production systems, although pig production industries do commonly utilize waste co- and by-products (e.g. tallows and oils or oilseed mills) as feed sources in diets. Some production systems are also designed to facilitate direct feeding of waste products from human food production systems, such as waste confectionary and dairy products. Thus, the pig production industry is well placed structurally and likely open to utilizing alternate feed protein sources such as insects. Improving sustainability in pig production systems requires a reduced footprint (e.g. by using more sustainable sources of feed, requiring less use of resources such as land or water) or a reduced input (feed choices and animal genetic selection pressures are based on using local feeds and selecting robust animals that can perform efficiently in a changing climate) or both. Insects can contribute to sustainability goals by both making less from more and by potentially improving animal growth efficiency and health. This chapter will review some of the current knowledge on the use of insects in pig diets, highlighting the current research trends and knowledge gaps. Where relevant, this chapter will highlight reviews and directions for further reading but will not cover the engineering or processing elements of insect production.

2 Why insects

Insects are rapidly growing, cold-blooded, require minimal space, and contribute to the natural diet of many species including pigs and poultry. Insects can bioconvert many organic (and potentially inorganic) substrates including vegetables and fruits, grains, manure, and animal remains into a product high in protein and lipids. While the feed conversion ratio of production animals ranges from 5.0 for pork (i.e. 1 kg meat gain requires 5 kg feed input) to 10.0 for beef, these ratios are approaching 2.0 for insects (van Huis, 2013). Fly larvae like black soldier flies (BSFL) and other insects can consume organic waste, sequester carbon, and provide a nutrient-rich biomass that can be consumed by animals or humans. One study found that bioconversion of waste released 70% less carbon into the atmosphere than composting (Perednia, 2017). Compared to commonly fed protein sources such as soybean meal, insects require considerably less land (and potentially water and energy) use. To produce a similar quantity of pork or chicken protein, enterprises would require 2–3.5 ha of land compared to only 1 ha of land to produce equivalent insect (mealworm) protein (van Huis et al., 2013).

Insect species being investigated for production as human feed ingredients include crickets, mealworms, and locusts while the animal feed sector is more focused on the production of BSFL, mealworms, silkworms, common houseflies, and crickets (van Huis, 2016). Of these, BSFLs appear to be the most suited for larger scale production and unlike other species, they are not disease vectors as they do not lay their eggs on decaying organic materials and the adults do not eat. The feed provided during rearing changes the chemistry of the resultant insects, which means the rearing substrate composition is imperative and can also be manipulated to somewhat specify the appropriate composition of the insects. For example, the fat content of BSFL was greatest in those reared on dairy products (DiGiacomo et al., 2019) followed by fruits and then vegetables (Jucker et al., 2017). While being a valuable source of high-quality protein and overall have an excellent balance of amino acids (Veldkamp and Vernooij, 2021), insects also provide a source of lipids, vitamins, and minerals but can also possess bactericidal, antifungal, antiviral, and antioxidant compounds that may be of use as nutraceutical additives in pig diets (Sogari et al., 2023).

3 Current state of insect production

Alternative protein sources being developed as animal feed are currently focused on insects, novel plant proteins, seaweed, algae, and other single-cell sources. Of these, insects are receiving global interest. While the commercial use of insects in pig diets is relatively new, pigs will naturally consume insects, particularly when dietary protein is scarce (Kierończyk et al., 2022). While the insect farming industry is in its infancy, large-scale research and investment are underway globally to confirm the safety, reliability, sustainability, and potential contribution of insects as an animal feed source. This increased interest in insects from a commercial perspective has been supported by an increase in academic research. For example, the European Association of Animal Production (EAAP) has been hosting sessions on insects in animal feed since 2017, and in 2019, the insect study commission was established.

The use of insects in pig diets is not a new idea, with the first publications examining this topic published in the late 1970s (Newton et al., 1977). There has been a steady increase in the number of publications in the last two decades which has been linked to an increase in research funding globally, and many of these publications are in the agricultural sector. According to Baiano (2020), 637 manuscripts with 'edible insect' in the title were published from 1975 until January 2020. In recent years, numerous reviews of the use of insects in pig diets have been published. I previously reviewed the role of insects in commercial pig diets (DiGiacomo and Leury, 2019) and summarized the findings for their use in pig diets. An updated version of these data is provided in Table 1, although this is by no means a complete list. In an invited review, Hong and Kim (2022) provide a comprehensive summary and reference list of current data from feeding insects to pigs, and this data will not be repeated in depth in this chapter. Given there is a large variation in the experimental methods including the type of insect and inclusion rate fed, age of pigs, number of animals fed, and nutrient content/processing of the insects, and the definitive conclusions regarding the use of insects as pig feed are not yet possible.

While Western countries are often averse to consuming insects, countries such as Papua New Guinea, Thailand, African countries, Mexico, India, China, and many others have traditionally consumed insects in human diets (DeFoliart, 1999). It is predicted that approx. 2 billion people consume insects regularly in over 130 countries (Costa-Neto and Dunkel, 2016). Indigenous populations, such as those in Mexico and Australia, consumed insects as a valuable and highly sought out food source, often associated with celebrations and special occasions (Meyer-Rochow and Changkija, 1997; DeFoliart, 1999). A significant barrier to the uptake of insects as human feed is the aversion to eating insects by Western consumers, likely driven by the association of insects with disease (Bartrim, 2017). Conversely, there is increased current interest in encouraging the consumption of insects in Western societies. As the flavor of insects changes with each species and can also be manipulated by the growth medium and preparation method, there is increased interest in the production of high-end insects for human consumption. For animal feed producers, the lack of uptake is primarily driven by the limited volumes and high prices of presently available products although fear and lack of understanding likely also contribute. Further, most farmers' previous experience with insects will be in dealing with insects as disease vectors or pests (or at least a nuisance), which makes it more difficult to change perceptions and acceptability of insects. According to one survey, farmers perceive insects in feed as having a better nutrient profile and improved sustainability but a lower microbiological safety (Verbeke et al., 2015).

The environmental and socioeconomic benefits of the insect industry may surpass current projections based solely on costs of input vs output given the wide variety of uses for insects and insect products. Insects can be farmed as bait, for medicinal purposes and for the provision of fertilizer. These properties are particularly beneficial to the production of small-scale farms in developing countries. Even when not feasible to grow insects in stand-alone farms, if used as additional income/product streams, there can be localized economic empowerment benefits, particularly for the development of food and economic security in developing nations. For example, producing insects in orphanages in the Democratic Republic of Congo can generate significant income (approx.

Table 1 Summary of some of the published growth responses in pigs fed diets containing insects

Animals used	Insect type fed	Inclusion rate	Production response	Reference
5-week-old barrows, n = 6 (Latin square)	BSFL	33% replacement of soybean meal	Increased feed intake ($P < 0.05$); reduced apparent DMD ($P < 0.05$)	Newton et al. (1977)
Early weaned pigs (n not reported)	BSFL	0%, 50%, or 100% replacement of dried plasma	50% diet improved performance; 100% diet decreased performance	Newton et al. (2005)
Castrated male piglets (n = 16); barrows (n = 16)	BSFL	Complete replacement of soybean meal with and without AA fortification	No effect on growth or intake; BSF combined with AA fortification improved piglet protein quality ($P < 0.005$); BSF improved apparent N digestibility	Neumann et al. (2018)
Barrows (n = 48)	BSFL	50%, 75%, and 100% replacement of soybean meal	No effect on base meat quality measures, increased juiciness ($P < 0.05$); higher backfat polyunsaturated fatty acid (PUFA) contents ($P < 0.05$)	Altmann et al. (2019)
Weaned female pigs (n = 48)	BSFL	0%, 30%, and 60% replacement of soybean meal	Linear increase in ADFI; no effect on growth	Biasato et al. (2019)
Crossbred female finishing pigs (n = 72)	BSFL	BSFL at 0%, 4%, or 8% of the diet for 46 days	4% BSFL reduced colonic *Streptococcus* and increased *Lactobacillus* ($P < 0.05$); increased short-chain fatty acids and decreased protein fermentation ($P < 0.05$); downregulated pro-inflammatory cytokines and upregulated anti-inflammatory cytokines ($P < 0.05$). 8% BSFL increased *Clostridium* clusters ($P < 0.05$) and increased butyrate ($P < 0.05$)	Yu et al. (2019)
Crossbred weaner pigs (n = 40)	BSFL	BSFL as a replacement of fish meal at 0%, 25%, 50%, 75%, or 100%	No effect of diet on feed intake, weight gain, or blood parameters except for platelet counts that were reduced in pigs fed 25%, 75%, or 100% BSFL ($P = 0.042$).	Chia et al. (2019)
Weaning piglets (n = 128)	BSFL	Full-fat BSF at 0%, 1%, 2%, or 4% (isoenergetic and isonitrogenous) over two feeding phases (1–14 days and 15–28 days)	Linear increase in BW, ADG, organ weights (liver, pancreas, small intestine) from 0 day to 14 days with BSFL inclusion ($P < 0.05$), linear and quadratic decrease in feed:gain and CP/CF digestibility ($P < 0.05$). Changes to metabolism and immune status and intestinal morphology	Yu et al. (2020)

(Continued)

Table 1 (*Continued*)

Animals used	Insect type fed	Inclusion rate	Production response	Reference
Growing pigs ($n = 10$, ileal cannulated)	BSFL and HFL	BSFL or houseflies (HF) as the primary source of nitrogen (vs. nitrogen-free diet) fed for 7 days. Titanium dioxide marker	The CSID of all AA for HF and BSFL were more than 0.726 and 0.641, respectively. The values for the CSID of all AA in HF were greater ($P < 0.05$) than in BSFL. The values for the CSID of all AA except methionine and cysteine in HF are greater ($P < 0.05$) than in BSFL	Tan et al. (2020)
Crossbred (Large White × Landrace) finisher pigs ($n = 40$)	BSFL	Fish meal replacement with BSFL at 0%, 25%, 50%, 75%, or 100% for 98 days	Carcass weight of pigs fed diets with BSFL replacing FM by 50%, 75%, or 100% (w/w) was higher ($P < 0.001$) than for pigs fed control diet with 100% FM as protein source. Mean FCR was decreased with diets containing ≥50% BSFL ($P < 0.05$). CP content of pork tissues was high (65–93% on dry-matter basis) across all dietary groups	Chia et al. (2021)
Castrated male pigs (ileal cannulated; $n = 8$)	BSFL	Latin square (four diets, five periods: nine replicates). Fish meal vs. defatted BSFL meal vs. adult BSF vs. nitrogen-free diet. Chromium oxide marker	The coefficient of standardized ileal digestibility (CSID) of CP in defatted BSFL meal was lower (0.738 vs. 0.883; $P < 0.05$) than that in fish meal but was higher ($P < 0.05$) than that in adult BSF (0.561). The CSID of all indispensable AA except for methionine and phenylalanine in defatted BSFL meal was lower ($P < 0.05$) than that in fish meal. The CSID of all AA except for proline in defatted BSFL meal was higher ($P < 0.05$) than that in adult BSF	Kim et al. (2023)
In vitro (pig GI tract)	Crickets, mealworms	Plant proteins vs. insects cooked four ways: raw, autoclaved (25 min), or oven cooked (150°C for 30 min or 200°C for 10 min)	Insect IVCPD decreased with thermal treatments ($P < 0.05$); mealworms were more digestible than crickets ($P < 0.01$).	Poelaert et al. (2016)
Castrated male piglets ($n = 21$)	Crickets	Three iso-nitrogenous diets (18.4% CP) including either fish meal (control), whole cricket meal, or body cricket meal (legs removed) fed *ad libitum*	Cricket diets improved DM intake ($P < 0.001$) and digestibility ($P < 0.05$), nutrient intake ($P < 0.001$), piglets were heavier ($P < 0.001$) and FCR ($P < 0.01$) was lower and N retention is greater ($P < 0.05$).	Miech et al. (2017)

Animal	Insect	Treatment	Results	Reference
Crossbred weaning pigs (n = 100)	Crickets	Replacement of fish meal with crickets on a lysine basis (100%), replacing fish meal a kg/kg basis (100%) or total replacement of soybean and fish meal for 28 days	Cricket supplementation decreased the incidence of diarrhea ($P < 0.05$), increased bodyweight and ADG ($P < 0.05$). Crickets increased CP digestibility ($P = 0.041$) and crude fat digestibility ($P = 0.024$). Crickets increased jejunal villus height ($P = 0.009$) and serum immunity markers ($P < 0.05$)	Boontiam et al. (2022)
Weaning pigs (n = 120)	Mealworm	Mealworm at 0%, 1.5%, 3%, 4.5%, or 6% inclusion (isoenergetic and isonitrogenous) for 35 days (2 phases, 0–14 days and 14–35 days)	Linear increase in intake, bodyweight, ADG during phase 1 ($P < 0.01$). During phase 2 ADG tended to increase ($P = 0.08$), increased ADG ($P < 0.01$) and feed intake ($P < 0.05$). Feed:gain tended to increase across the whole period ($P = 0.07$). Linear increase in N retention and DMD and digestibility of CP ($P = 0.05$). Linear decrease in blood urea N ($P = 0.05$) and increase in IGF-1 ($P = 0.03$)	Jin et al. (2016)
Crossbred weaned pigs (n = 240)	Mealworm	Fish meal portion of the diet replaced at 0%, 50%, or 100% for 35 days	100% mealworm increased feed:gain during the first week ($P < 0.05$). During the second week, 50% mealworm increased ADG ($P < 0.05$), although across the whole period 50% replacement decreased ADG and final BW ($P < 0.05$)	Ao et al. (2020)
Weaning pigs (n = 180)	Mealworm	Fish meal portion of the diet replaced at 0%, 50%, or 100% for 28 days (two phases, 0–14 days and 15–28 days).	Feed to gain was increased in the 100% compared to 50% treatment in phase 1 ($P = 0.024$). The ADG was greater in 50% compared to 100% treatments ($P = 0.01$). Plasma IgG increased with mealworm diets in phase 1 ($P = 0.048$).	Ko et al. (2020)
Male 5-week-old, crossbred pigs (n = 30)	Mealworm	Mealworm at 0%, 5%, or 10% inclusion (isoenergetic and isonitrogenous) for 4 weeks	Decrease in AA digestibility ($P < 0.05$), weak impact on intermediary metabolism (using omics analysis)	Meyer et al. (2020)
Crossbred weaned pigs (n = 180)	Mealworm	Mealworm at 0%, 1%, or 2% in the diet (replacement of fish meal)	Pigs fed 1% mealworm had reduced final BW compared to control ($P < 0.05$). During days 0–7, feed:gain ratio was higher in mealworm diets ($P<0.05$). Feeding 1% mealworm decreased ($P < 0.05$) ADG during days 8–21 and 0–35. The 1% mealworm treatment had lower apparent total tract digestibility of DM and nitrogen ($P < 0.05$)	Ao and Kim (2019)

(Continued)

Table 1 (*Continued*)

Animals used	Insect type fed	Inclusion rate	Production response	Reference
Piglets (n = 144)	Mealworm, HFL, superworms	(Companion papers) Soybean-based diet with 5% of plasma protein powder (control), 5% mealworm powder, 5% HFL powder, or 5% superworm powder. Fed in two phases (1–28 days and 29–56 days)	ADFI was decreased by mealworm and HFL on day 7 and increased in superworm diets on days 28 and 56 ($P < 0.05$). Insect diets decreased diarrhea from days 15–28 ($P < 0.05$) and decreased plasma ammonia ($P < 0.05$) and increased the methionine AID ($P < 0.05$)	Ji et al. (2016)
	Mealworm, HFL, superworms	(As earlier, companion papers)	Plasma lysine concentrations were reduced in mealworm and superworm diets during phase 1 ($P < 0.05$). Insect powders overall altered AA transporters and improved the metabolism of AAs	Liu et al. (2020)

Please note, this list is not reflective of all current published data.

AA: amino acid; ADG: Average daily gain; ADFI: Average daily feed intake; AID: apparent ileal digestibility; BSFL: Black soldier fly larvae; BW: body weight; CF: crude fat; CP: crude protein; CSID: coefficient of standardized ileal digestibility; DM: dry matter; DMD: dry matter digestibility; FCR: feed conversion ratio; HFL: housefly larvae; IgG: immunoglobulin G; IGF-1: insulin-like growth factor 1.

$300 US/month) while also producing a sustainable and nutritious food and feed source (Franklin et al., 2018).

4 Potential insects and current production volumes

When selecting suitable insects, producers need to be sure they do not contain toxins, venoms, or unpalatable exoskeletons. Selection also needs to ensure that risks are avoided or minimized. For example, mealworms are a pest of grains and flower products, while houseflies are a pest to humans and other animals. There is a large variety of insects that can be reared for animal and human consumption. Those most investigated to date include crickets, flies (black soldier and house flies), silkworms, crickets, mealworms (giant, lesser, etc.), moths (wax moths, etc.), locusts, and grasshoppers. The selection of an appropriate insect species for individual production systems or target consumers will need to consider multiple factors including feed substrate available, environmental conditions, facilities, reproductive rate, size, behavior, disease risk, storage potential, and marketability. The lifecycle of insects will differ between species and some insects are consumed during their larval phase (i.e. flies), while others are consumed as adults (i.e. crickets). Thus, the length of time and input required to rear and farm these insects will be highly varied as it is when comparing the rearing of pigs to beef cows for example.

In 2019, Mancuso et al. (2019) reported that the European insect industry was producing low volumes of around 3000 tonnes of protein and BSFs were retailing for €2-9 per kg and mealworms retailing for €10-32 per kg. While these prices are not current, they do demonstrate the high cost of the protein driven by the low production capacity of the industry. It is predicted that the European industry will be producing 1.2 million tonnes by 2025 (IPIFF, 2019), although the COVID-19 pandemic may have stalled some growth in the sector. An increase in production volume should see a reduction in prices, but this will depend upon input costs and market pressures. Production may begin to increase more rapidly given that regulatory changes occurring in 2021 approved 'processed insect protein' for incorporation into animal diets (including pigs) in Europe. This is a positive step forward given that in the EU insects were previously prohibited from use in animal feed based on the risks (real or perceived) associated with transmissible spongiform encephalopathy (TSE).

Currently most insect production companies remain in the start-up phase and are not yet producing at a large or consistent scale. This lack of scale is driven by numerous factors including technology and engineering controls, access to capital investment, lack of customer base given the small-scale volumes presently produced, and lack of knowledge or research specific to each production system and environment. Given most insect producers were motivated to enter the industry because of their interest in contributing to sustainability goals, and while engineering and processing systems remain under development, most producers are attempting to develop environmentally sustainable processing methods utilizing existing efficient technologies such as solar power. While this review will not discuss the technical processes of rearing insects on a commercial scale, rearing processes must be developed for efficiency in output, energy/resource usage, and greenhouse gas emissions to remain sustainable. For example, as rearing temperature can impact the conversion efficiency of insects, energy inputs may be required to maintain appropriate growth climates which would increase energy resource usage (Rumpold and Schlüter, 2013). Further, given the scale of feed required for global pig diets, it is unlikely that insects would be more than just a sidestream or small diet component rather than a major constituent or complete replacement of traditional protein sources (such as soybean), but even small improvements in sustainability, efficiency, and reductions in costs could be valuable.

5 Production and health responses in pigs

Research into the feeding of insects to commercial pigs (from piglets to finishing pigs) has increased in recent years. Numerous reviews have been published summarizing such findings and thus will not be repeated here (e.g. Hong and Kim, 2022; Veldkamp and Bosch, 2015; Veldkamp et al., 2022). In a comprehensive recent review, Hong and Kim (2022) summarized the current literature examining insect digestibility and the impacts on meat quality in pork production systems, citing >20 publications from across the globe since 2020. Overall, including insects in pig diets either improved or did not alter growth rates, health, and meat quality. For example, weaning piglets had increased weight and efficiency markers when fed BSFL (Yu et al., 2020), crickets (Boontiam et al., 2022), or mealworms (Jin et al., 2016). In barrows, feeding BSFL had no negative effect on meat quality when fed as a replacement for soybean meal (Altmann et al., 2019). Recently, Ipema et al. (2021) examined the use of live BSFL as feed and enrichment for pigs and showed that the BSFL was preferred compared to corn, pellets, or raisins. Pigs not only interacted with the feed dispersal mechanisms but also consumed all the BSFL (Ipema et al., 2021), indicating that the provision of live insects can not only contribute nutrients but also enrich and the promotion of a positive welfare state.

Although the published literature is inconsistent with experimental design and diet composition, the consensus is that insect protein nutrient digestibility is comparable with traditional protein sources. It is important to note that digestibility of house crickets and mealworm larvae decreased when comparing raw to heated products (oven baked at either 150°C or 200°C or autoclaved) in pigs using an *in vitro* model (Poelaert et al., 2016), which is the opposite to most grains which increase in digestibility when cooked. Antinutritional factors, including chitin, can also decrease the digestibility of insects when fed to pigs while the addition of a protease can reduce some of these negative responses (Go et al., 2022). The exoskeletons of insects can be rich in chitin, which can therefore decrease digestibility. Given this, the analysis of insect protein concentrations using the Kjeldahal method overestimates the available protein (Jonas-Levi and Martinez, 2017). However, chitin and its derivatives can improve immunity, act as an antioxidant and antimicrobial, and can therefore improve growth performance in growing animals (Swiatkiewicz et al., 2015). Despite this potential decline in digestibility, most studies have shown either no change or an improvement in growth performance when feeding insects. However, without optimization of rearing and processing methods and inclusion rates, it is difficult to draw complete conclusions.

6 Competition for resources

A major initiative to reduce food waste will require cooperation between producers and consumers to both reduce oversupply and efficiently utilize unavoidable waste (FAO, 2011). Insects reared by and for the pig production industry can certainly contribute to this process. While most of the published literature has focused on the specifics of insect production including nutrient content, safety, and processing practice; a few have examined the economics and production volume capabilities. Substrates investigated for rearing insects include food and feed waste, manure, horticulture waste, and other organic substrates. Forest or gardening waste could also be utilized as rearing substrates if logistical difficulties could be overcome. For example, working in conjunction with local governments to redirect organic waste from landfills or even build insect processing plants within waste processing sites. Substrates could also include agricultural waste products, such as plant stalks and leaves, condemned grains (mycotoxins, etc.), manure, hulls and skins, spent bedding, etc. Many of these substrates can be in competition with other production systems including direct competition with pork production systems who already utilize waste products in diets as much as possible. However, in most cases processing waste incurs a cost (labor, transport, disposal fees, etc.). Insects like BSF consume waste rapidly and they generally consume the most digestible portion of the substrate first, meaning valuable nutrients can be left unchanged in the unconsumed waste. This leftover feed together with insect waste (excrement and exoskeletons for example) is called frass and is a nutrient-rich fertilizer product. In sub-Saharan Africa where soils are mostly nutrient deficient, including BSFL frass as a fertilizer increased the income from the farming practice (Beesigamukama et al., 2022). Considering that for every kilogram of larvae produced 10-34 kg of frass is produced (Beesigamukama et al., 2022), frass can become a valuable income stream to accompany the insects themselves.

While some feed protein sources such as meat meals and tallow already efficiently utilize waste products from the animal production industries, they require processing to be suitable for use in monogastric animal diets. Like insect processing, drying and transport processes are likely to be the greatest contributors to energy use (and hence emissions). Like most livestock industries, the pig industry has little to no control of grain feed prices, so sourcing local and sustainably produced and price-stable alternatives is critical for industry longevity and profitability. In addition, organic waste remains a sizable problem to be addressed. For example, insects could be reared on waste or contaminated animal feed and animal bedding that would otherwise be sent to end-point waste systems like ponds and pits, contributing to a no-waste system and potentially providing additional income streams for the piggery.

Pigs, poultry, companion animals, and humans are the most suited target species for utilizing insect protein. Given the current high cost of insect protein, the more likely purchasers are going to be the industries with a higher margin and thus more money to spend on raw ingredients. Thus, those with tighter margins such as poultry and pig production systems may be unlikely to afford insect protein in its current capacity and price. Currently, there is increased interest in sourcing insects for use in companion animal diets who have larger margins and a greater ability to spend on feed ingredients, given consumers are prepared to pay premium prices for pet food products. There is a global increase in the number of producers manufacturing pet feed containing insects, including Mars Petcare in the UK who are selling an insect-based cat food 'Lovebug'. Similarly, Nestlé Purina is producing insect and plant protein-based pet food in Switzerland.

Previous research has shown that insects (BSFL and mealworms, for example) can consume feed contaminated with mycotoxins without becoming contaminated themselves (Purschke et al., 2017; van der Fels-Klerx et al., 2018). While the metabolic processes the insects utilize to achieve this remain unknown, it is an exciting possibility that insects could consume contaminated feed, thereby utilizing waste feed and also providing a safe insect product that can be utilized as feed or food (Bosch et al., 2017; Gützkow et al., 2021; Scheibelberger et al., 2017). Growing BSFL did not accumulate pesticides when fed spiked feed (Cai et al., 2018b) although heavy metals (Cd, Cr, Ni, Pb, Hg, and As) are accumulated in the BSFL in significant quantities in multiple studies (Biancarosa et al., 2018; Cai et al., 2018a). In a comprehensive review, Lievens et al. (2021) summarize the current knowledge on the accumulation of toxins and heavy metals in insects, concluding that while investigations to date have shown no negative impact on insect growth when fed toxins or heavy metals, producers, and consumers need to ensure the product safety by monitoring any potential accumulation.

Recent evidence suggests that BSFL may be able to consume microplastics (from plastics commonly found in disposable bags and containers) (Cho et al., 2020) and polystyrene waste (Peng et al., 2019). Australian research showed that feeding some types of plastic did not diminish BSFL or mealworm growth and survival rates (although porous polylactic acid blocks reduced BSFL weight by ~29%) (Beale et al., 2022). This is promising as it could increase the options for feed sources, provide a useful mechanism to reduce some of the plastic waste ending up in landfill, and potentially reduce processing requirements if some plastic could be included in insect diets (reducing the need to sort or unpack waste).

7   Rearing insects on manure

Numerous insects can consume (and thrive on) manure, including BSFL, mealworms, and houseflies, as reviewed by Cammack et al. (2021). This can assist in the processing of waste that may not always be utilized (e.g. if biogas facilities are not present on farms). In addition, the residue from corn cob fermentation used to produce biogas has been effectively used in conjunction with pig manure as a substrate for rearing BSFL (Li et al., 2015). Thus, biogas residues, that are commonly used as fertilizer, can potentially be bioconverted into a more valuable product resulting in both frass fertilizer and proteins/oils from insects. Utilizing manure as a rearing substrate may be a useful approach to utilize substrates in a meaningful way that would otherwise contribute to the pollution of soils and waterways (via runoff and leaching) and the production of greenhouse gases. This process can generate useful products such as animal feed, oils for biodiesel and cosmetics, protein for use as emulsifiers and bioplastic applications, and chitin for use in biomedical applications and frass fertilizer. Previous experiments have demonstrated that BSFL can reduce the dry weight of manure by 30-50% and reduce the nitrogen content by 30-80% (Myers et al., 2008; Newton et al., 2005; Oonincx et al., 2015). Pig manure is a suitable substrate for mealworms, beetles (*Protaetia brevitarsis sulensis*), and flies (*P. tenebrifer*); each recording an approx. 90% survival rate (Choi, 2022). Up to a 100% reduction (ranging 87-100%) in volatile organic compounds have been recorded when BSFLs are reared on manure (Beskin et al., 2018). Insects such as BSFL can also reduce the pathogenic bacterial count of pig manure (Elhag et al., 2018), the *E. coli* counts in dairy manure (Liu et al., 2008), and *E. coli* and *Salmonella enterica* counts in dairy and chicken manure (Erickson et al., 2004). Processing manure with BSFL not only creates pathogen-free fertilizer products for farm use - far safer than directly adding pathogen-rich manures to soils - but can potentially also turn manure into livestock feed.

Using manure as a substrate doesn't come without risks, the greatest of which is the risk of contamination from veterinary medicines. However, Charlton et al. (2015) showed that larvae fed with manure containing veterinary medicines generally did not accumulate the substances (except for domestic flies) and most medications tested in the larvae were below the limit of detection. However, spreading contaminated manure would also be a potential source of risk to environmental contamination, particularly for residues in soil and water. In another experiment, BSFL were shown to degrade but not accumulate three pharmaceutical compounds (carbamazepine, roxithromycin, and trimethoprim) (Lalander et al., 2016), while tetracycline was degraded more efficiently by BSFL than normal composting (Cai et al., 2018b). Thus, processing manure using insects may prove to be the most environmentally friendly option. The use of manure as a rearing substrate for insects to be fed to production animals is currently prohibited in the EU and there will likely be resistance to this changing given insect's current classification as farmed animals (Boloh, 2018). Additional research in this area may assist to persuade change, particularly when evidence suggests that BSF larvae can reduce the pathogenic bacterial counts.

8 Current research

8.1 Insects as antimicrobials

Using insects as a source of antimicrobials has received increased attention as a potential natural antibiotic alternative to use as a disease prevention and growth-promoting agent. In a review, Jozefiak and Engberg (2017) concluded that antimicrobial peptides sourced from insects have excellent potential to be effectively used in animal diets. The combined impacts of increased demand, increased pressure to improve efficiency, and a desire to reduce antibiotic use and prevent further antimicrobial resistance have emphasized the urgent need to find alternative agents. Using insects as a feed and food source of microbial may be a valuable contribution, especially given there is evidence that antimicrobial peptides (AMPs) from insects do not result in the development of resistance (Jozefiak and Engberg, 2017). As reviewed by Hadj Saadoun et al. (2022), insects should be further examined as sources of antimicrobials, and it is likely that the estimated amount of AMP in insects has been underreported to date.

8.2 Energy use and life cycle analysis

As described previously, insect-rearing processes require labor and energy inputs similar to any farm enterprise. Life cycle assessments seek to quantify total emissions from inputs (such as animal feed) to processing (such as drying) and transport. van Zanten et al. (2018) concluded that feeding BSFL as a replacement for soybean meal in pig diets reduced global warming potential and land use, though life cycle assessment demonstrated no change in global warming potential. The complexities of diet formulations, particularly for pig and poultry diets that often include multiple ingredients, mean that substitution feed products with insects or alternate proteins will not always result in a net reduction in overall environmental impact. This is due to the many steps required such as land and water use, transport, processing, storage, and so on. While water use by insect rearing processes cannot yet be fully assessed given processing methods are under development, larvae and are predicted to have lower water, and certainly land use, compared to terrestrial animals and crops.

9 Future trends in research

9.1 Alternate insect species

Native species in individual countries will potentially be more efficient farmed insect targets given they have evolved to suit local climates. Future research should involve investigating a wider range of insect species targeted to specific regions. Further, harvesting pest species from agricultural crops etc. would be an ideal scenario as it would utilize the pests in a meaningful way while also improving existing production. However, the ability to capture such pests in an efficient manner could make this concept quite complicated but likely solvable with investments into research and development of harvesting technologies.

9.2 Insect welfare

The welfare of insects is often overlooked or dismissed as unimportant in this emerging industry. Yet, if insects are classified as farmed animals, then their welfare will need to be considered. In a recent review, Lambert et al. (2021) demonstrated that there is significant scientific evidence to support the notion that insects (across a range of orders) display sentience and cognition and are also likely to feel stress and pain. Thus, the development of farming systems to produce insects on a large scale must undertake thorough research to ensure their harvesting and slaughter practices are humane and ethical. To the best of my knowledge, no such research has been undertaken to date.

9.3 Impact of coronavirus disease 2019 on global production and feed sources

The COVID-19 pandemic impacted global trade, which of course impacted the pork production industries by increasing the costs of feed freight. Feed supplies were therefore more expensive and when combined with a reduction in pork meat product prices, the pork industry profit margins for pork were reduced. Given that regular animal production and processing chains were disrupted, the global market for animal feed protein ingredients was impacted. For example, as summarized by Rahimi et al. (2022), soybean meal exports from the Americas were reduced during the pandemic. Future events such as pandemics and weather events are likely to continue to disrupt the feed supply chain, further bolstering the need for sustainable and adaptable locally produced animal feed ingredients which may include insects.

9.4 Workplace labor constraints

The insect production industry is developing automated and robotic-based systems and engineering controls to reduce the need for manual labor. While such technological advancements will ultimately improve processing and accuracy, it will not eliminate the need for labor and will shift the type of jobs required to be more control and data based rather than manual labor. Therefore, training and advancement of employees in more technological based skills (such as operating robots and computer based systems) is required to ensure a sustainable and diverse workforce can be maintained for the insect production industry. Given the relative infancy of the industry, there is a timely opportunity to target and involve workers from a diversity of backgrounds (age, location, cultural background, gender, etc.), attract a wide range of skills (e.g. digital/precision agriculture), and attract staff interested in contributing towards a more sustainable agricultural industry.

10   Barriers to uptake/challenges to production

The greatest current challenge to insect production is the lack of scale. Engineering, infrastructure, and market processes need to be developed and optimized. In an opinion piece, Aarts (2019) describes the complexities of starting an insect rearing business (based on their experience with Protix) and suggests that insect producers need to consider the science, innovate appropriately, realize that insects are not the only potential source of alternate proteins, and importantly collaborative efforts need to occur in all facets of the system. A lack of collaboration between producers is currently constraining industry advancement. This is mostly due to perceived and real risks associated with non-mutual benefits and commercial competitiveness (Ponce-Reyes and Lessard, 2021). That being said, there are numerous large-scale insect production research programs bringing together academia and industry globally, such as the insect value chain in a circular bioeconomy (inVALUABLE) in Europe, involving 10 partners and worth over €3.7 million (Heckmann, 2018). Continued investment in collaborative research projects (across insect producers, states, universities, and agricultural and government industries) is imperative.

Consumer education will also be paramount if insects are going to become a more major and consistent feed (or food). In a 2018 survey, Weinrich and Busch (2021) identified that most consumers did not feel well educated about animal feed, although negative responses towards soybean and GM feeds are often expressed. However, consumers are increasingly concerned with the environmental impact of their food choices. Highlighting the environmental and sustainability benefits of insect production will likely lead to positive consumer responses if the messaging is clear and supported by evidence.

11 Conclusion

This chapter has highlighted some of the major findings and knowledge gaps in the insect production industry. Pork production industries have the opportunity to adopt the use of insects for inclusion as a feed source but also as a tool to process waste. Insects can provide valuable nutrients to pigs including fat, protein, and antimicrobials. Despite significant investment into the upscaling of insect production, the industry remains in its infancy and extensive adoption is yet to occur. Further research is required to establish and optimize insect rearing, storage, and processing practices and develop new markets; to establish insect production emissions and energy use values; to confirm insect inclusion levels and product safety; to examine the safety of insects reared on manure; and to confirm the potential bioactive and other functional properties of insects. Pig production industries are well placed to contribute to the reduction of waste and the improvement of the sustainability of animal production industries globally by embracing the use of insects as feed and as a waste management activity.

12 Where to look for further information

Numerous reviews and book chapters on the use of insects in the production of animal diets have been published in recent years. Some suggestions are provided below, although this list is by no means exhaustive.

- Veldkamp, T. and Vernooij, A. G. (2021). Use of insect products in pig diets. *Journal of Insects as Food and Feed*, 7(5), 781–793. doi:10.3920/jiff2020.0091.
- Hong, J. and Kim, Y. Y. (2022). Insect as feed ingredients for pigs. *Animal Bioscience*, 35(2), 347–355. doi:10.5713/ab.21.0475.
- van Huis, A. and Tomberlin, J. K. (Eds.). (2017). *Insects as Food and Feed. From Production to Consumption.* https://doi.org/10.3920/978-90-8686-849-0.

- Parrini, S., Aquilani, C., Pugliese, C., Bozzi, R. and Sirtori, F. (2023). Soybean Replacement by alternative protein sources in pig nutrition and its effect on meat quality. *Animals*, 13(3), 494. doi:10.3390/ani13030494.
- Lu, S., Taethaisong, N., Meethip, W., Surakhunthod, J., Sinpru, B., Sroichak, T., Paengkoum, P. (2022). Nutritional composition of black soldier fly larvae (*Hermetia illucens* L.) and its potential uses as alternative protein sources in animal diets: a review. *Insects*, 13(9). doi:10.3390/insects13090831.

13   References

Aarts, K. (2019). How to develop insect-based ingredients for feed and food? A company's perspective. *Journal of Insects as Food and Feed* 6(1), 67-68.

Altmann, B. A., Neumann, C., Rothstein, S., Liebert, F. and Mörlein, D. (2019). Do dietary soy alternatives lead to pork quality improvements or drawbacks? A look into micro-alga and insect protein in swine diets. *Meat Science* 153, 26-34.

Ao, X. and Kim, I. H. (2019). Effects of dietary dried mealworm (*Ptecticus tenebrifer*) larvae on growth performance and nutrient digestibility in weaning pigs. *Livestock Science* 230, 103815.

Ao, X., Yoo, J. S., Wu, Z. L. and Kim, I. H. (2020). Can dried mealworm (*Tenebrio molitor*) larvae replace fish meal in weaned pigs? *Livestock Science* 239, 104103.

Baiano, A. (2020). Edible insects: an overview on nutritional characteristics, safety, farming, production technologies, regulatory framework, and socio-economic and ethical implications. *Trends in Food Science and Technology* 100, 35-50.

Bartrim, J. (2017). Insect farming and consumption in Australia - opportunities and barriers. *Australian Zoologist* 39(1), 26-30.

Beale, D. J., Shah, R. M., Marcora, A., Hulthen, A., Karpe, A. V., Pham, K., Wijffels, G. and Paull, C. (2022). Is there any biological insight (or respite) for insects exposed to plastics? Measuring the impact on an insects central carbon metabolism when exposed to a plastic feed substrate. *Science of the Total Environment* 831, 154840.

Beesigamukama, D., Mochoge, B., Korir, N., Menale, K., Muriithi, B., Kidoido, M., Kirscht, H., Diiro, G., Ghemoh, C. J., Sevgan, S., Nakimbugwe, D., Musyoka, M. W., Ekesi, S. and Tanga, C. M. (2022). Economic and ecological values of frass fertiliser from black soldier fly agro-industrial waste processing. *Journal of Insects as Food and Feed* 8(3), 245-254.

Beskin, K. V., Holcomb, C. D., Cammack, J. A., Crippen, T. L., Knap, A. H., Sweet, S. T. and Tomberlin, J. K. (2018). Larval digestion of different manure types by the black soldier fly (Diptera: Stratiomyidae) impacts associated volatile emissions. *Waste Management* 74, 213-220.

Biancarosa, I., Liland, N. S., Biemans, D., Araujo, P., Bruckner, C. G., Waagbø, R., Torstensen, B. E., Lock, E. J. and Amlund, H. (2018). Uptake of heavy metals and arsenic in black soldier fly (*Hermetia illucens*) larvae grown on seaweed-enriched media. *Journal of the Science of Food and Agriculture* 98(6), 2176-2183.

Biasato, I., Renna, M., Gai, F., Dabbou, S., Meneguz, M., Perona, G., Martinez, S., Lajusticia, A. C. B., Bergagna, S., Sardi, L., Capucchio, M. T., Bressan, E., Dama, A., Schiavone, A. and Gasco, L. (2019). Partially defatted black soldier fly larva meal inclusion in piglet diets: effects on the growth performance, nutrient digestibility, blood profile, gut morphology and histological features. *Journal of Animal Science and Biotechnology* 10, 12.

Boloh, Y. (2018). Insect proteins inch toward approval for EU animal feed: after the authorization for insect proteins in aquafeed in July 2017, Europe may accept insect proteins for poultry and pig feeds in 2019. *Feed Strategy* 69, 16.

Boontiam, W., Hong, J., Kitipongpysan, S. and Wattanachai, S. (2022). Full-fat field cricket (*Gryllus bimaculatus*) as a substitute for fish meal and soybean meal for weaning piglets: effects on growth performance, intestinal health, and redox status. *Journal of Animal Science* 100(4), skac080.

Bosch, G., Fels-Klerx, H. J. V., Rijk, T. C. and Oonincx, D. G. A. B. (2017). Aflatoxin B1 tolerance and accumulation in black soldier fly larvae (*Hermetia illucens*) and yellow mealworms (*Tenebrio molitor*). *Toxins (Basel)* 9(6), 185.

Cai, M., Hu, R., Zhang, K., Ma, S., Zheng, L., Yu, Z. and Zhang, J. (2018a). Resistance of black soldier fly (Diptera: Stratiomyidae) larvae to combined heavy metals and potential application in municipal sewage sludge treatment. *Environmental Science and Pollution Research* 25(2), 1559–1567.

Cai, M., Ma, S., Hu, R., Tomberlin, J. K., Yu, C., Huang, Y., Zhan, S., Li, W., Zheng, L., Yu, Z. and Zhang, J. (2018b). Systematic characterization and proposed pathway of tetracycline degradation in solid waste treatment by *Hermetia illucens* with intestinal microbiota. *Environmental Pollution* 242(A), 634–642.

Cammack, J. A., Miranda, C. D., Jordan, H. R. and Tomberlin, J. K. (2021). Upcycling of manure with insects: current and future prospects. *Journal of Insects as Food and Feed* 7(5), 605–619.

Charlton, A. J., Dickinson, M., Wakefield, M. E., Fitches, E., Kenis, M., Han, R., Zhu, F., Kone, N., Grant, M., Devic, E., Bruggeman, G., Prior, R. and Smith, R. (2015). Exploring the chemical safety of fly larvae as a source of protein for animal feed. *Journal of Insects as Food and Feed* 1(1), 7–16.

Chia, S. Y., Tanga, C. M., Osuga, I. M., Alaru, A. O., Mwangi, D. M., Githinji, M., Dubois, T., Ekesi, S., Van Loon, J. J. A. and Dicke, M. (2021). Black soldier fly larval meal in feed enhances growth performance, carcass yield and meat quality of finishing pigs. *Journal of Insects as Food and Feed* 7(4), 433–447.

Chia, S. Y., Tanga, C. M., Osuga, I. M., Alaru, A. O., Mwangi, D. M., Githinji, M., Subramanian, S., Fiaboe, K. K. M., Ekesi, S., Van Loon, J. J. A. and Dicke, M. (2019). Effect of dietary replacement of fishmeal by insect meal on growth performance, blood profiles and economics of growing pigs in Kenya. *Animals* 9(10), 705.

Cho, S., Kim, C.-H., Kim, M.-J. and Chung, H. (2020). Effects of microplastics and salinity on food waste processing by black soldier fly (*Hermetia illucens*) larvae. *Journal of Ecology and Environment* 44(1), 7.

Choi, I. H. (2022). Decomposition abilities and characteristics of pig manure using three insect larvae. *Entomological Research* 52(10), 439–444.

Costa-Neto, E. M. and Dunkel, F. V. (2016). Chapter 2. Insects as food: history, culture, and modern use around the world. In: Dossey, A. T., Morales-Ramos, J. A. and Rojas, M. G. (Eds). *Insects as Sustainable Food Ingredients*. Academic Press, San Diego.

Defoliart, G. R. (1999). Insects as food: why the western attitude is important. *Annual Review of Entomology* 44, 21.

Digiacomo, K., Akit, H. and Leury, B. J. (2019). Insects: a novel animal-feed protein source for the Australian market. *Animal Production Science* 59(11), 2037–2045.

Digiacomo, K. and Leury, B. J. (2019). Review: insect meal: a future source of protein feed for pigs? *Animal* 13(12), 3022-3030.

Elhag, O. A. O., Xiao, X. P., Zheng, L. Y. and Zhang, J. B. (2018). Antibacterial activity of *Hermetia illucens* against pathogen naturally present in the pig manure and its mechanism. The 2nd International Conference 'Insects to Feed the World', 15-18 May 2018, Wuhan, China. Wageningen Academic Publishers.

Erickson, M. C., Islam, M., Sheppard, C., Liao, J. and Doyle, M. P. (2004). Reduction of Escherichia coli O157:H7 and *Salmonella enterica* serovar enteritidis in chicken manure by larvae of the black soldier fly. *Journal of Food Protection* 67(4), 685-690.

FAO (2011). *Global Food Losses and Food Waste - Extent, Causes and Prevention*. FAO, Rome, Italy.

Franklin, A., Brandt, N. and Ureda, N. (2018). Edible insect farming: a strategy for providing sustainable nutrition and economic empowerment for orphanages in the DRC. The 2nd International Conference 'Insects to Feed the World' 15-18 May 2018, Wuhan, China. Wageningen Academic Publishers.

Go, Y. B., Lee, J. H., Lee, B. K., Oh, H. J., Kim, Y. J., An, J. W., Chang, S. Y., Song, D. C., Cho, H. A., Park, H. R., Chun, J. Y. and Cho, J. H. (2022). Effect of insect protein and protease on growth performance, blood profiles, fecal microflora and gas emission in growing pig. *Journal of Animal Science and Technology* 64(6), 1063-1076.

Gützkow, K. L., Ebmeyer, J., Kröncke, N., Kampschulte, N., Böhmert, L., Schöne, C., Schebb, N. H., Benning, R., Braeuning, A. and Maul, R. (2021). Metabolic fate and toxicity reduction of aflatoxin B1 after uptake by edible *Tenebrio molitor* larvae. *Food and Chemical Toxicology* 155, 112375.

Hadj Saadoun, J., Sogari, G., Bernini, V., Camorali, C., Rossi, F., Neviani, E. and Lazzi, C. (2022). A critical review of intrinsic and extrinsic antimicrobial properties of insects. *Trends in Food Science and Technology* 122, 40-48.

Heckmann, L. H. (2018). Research and development efforts on optimizing key parameters in industrial insect production. The 69th Annual Meeting of the European Federation of Animal Science, Dubrovnik, Croatia. Wageningen Academic.

Hong, J. and Kim, Y. Y. (2022). Insect as feed ingredients for pigs. *Animal Bioscience* 35(2), 347-355.

Ipema, A. F., Gerrits, W. J. J., Bokkers, E. A. M., Kemp, B. and Bolhuis, J. E. (2021). Live black soldier fly larvae (*Hermetia illucens*) provisioning is a promising environmental enrichment for pigs as indicated by feed- and enrichment-preference tests. *Applied Animal Behaviour Science* 244, 105481.

IPIFF (2019). *The European Insect Sector Today: Challenges, Opportunities and Regulatory Landscape. IPIFF Vision Paper on the Future of the Insect Sector Towards 2030*. IPIFF, Brussels.

Ji, Y. J., Liu, H. N., Kong, X. F., Blachier, F., Geng, M. M., Liu, Y. Y. and Yin, Y. L. (2016). Use of insect powder as a source of dietary protein in early-weaned piglets. *Journal of Animal Science* 94, 111-116.

Jin, X. H., Heo, P. S., Hong, J. S., Kim, N. J. and Kim, Y. Y. (2016). Supplementation of dried mealworm (*Tenebrio molitor* larva) on growth performance, nutrient digestibility and blood profiles in weaning pigs. *Asian-Australasian Journal of Animal Sciences* 29(7), 979-986.

Jonas-Levi, A. and Martinez, J.-J. I. (2017). The high level of protein content reported in insects for food and feed is overestimated. *Journal of Food Composition and Analysis* 62, 184-188.

Jozefiak, A. and Engberg, R. M. (2017). Insect proteins as a potential source of antimicrobial peptides in livestock production: a review. *Journal of Animal and Feed Sciences* 26(2), 87.

Jucker, C., Erba, D., Leonardi, M. G., Lupi, D. and Savoldelli, S. (2017). Assessment of vegetable and fruit substrates as potential rearing media for *Hermetia illucens* (Diptera: Stratiomyidae) larvae. *Environmental Entomology* 46(6), 1415–1423.

Kierończyk, B., Rawski, M., Mikołajczak, Z., Homska, N., Jankowski, J., Ognik, K., Józefiak, A., Mazurkiewicz, J. and Józefiak, D. (2022). Available for millions of years but discovered through the last decade: insects as a source of nutrients and energy in animal diets. *Animal Nutrition* 11, 60–79.

Kim, J., Ji, S. Y. and Kim, B. G. (2023). Ileal digestibility of amino acids in defatted black soldier fly larva meal and adult black soldier fly fed to pigs. *Journal of Insects as Food and Feed*, in press, 9(10), 1345–1352.

Ko, H., Kim, Y. and Kim, J. (2020). The produced mealworm meal through organic wastes as a sustainable protein source for weanling pigs. *Journal of Animal Science and Technology* 62(3), 365–373.

Lalander, C., Senecal, J., Gros Calvo, M., Ahrens, L., Josefsson, S., Wiberg, K. and Vinnerås, B. (2016). Fate of pharmaceuticals and pesticides in fly larvae composting. *Science of the Total Environment* 565, 279–286.

Lambert, H., Elwin, A. and D'cruze, N. (2021). Wouldn't hurt a fly? A review of insect cognition and sentience in relation to their use as food and feed. *Applied Animal Behaviour Science* 243, 105432.

Li, W., Li, Q., Zheng, L., Wang, Y., Zhang, J., Yu, Z. and Zhang, Y. (2015). Potential biodiesel and biogas production from corncob by anaerobic fermentation and black soldier fly. *Bioresource Technology* 194, 276–282.

Lievens, S., Poma, G., De Smet, J., Van Campenhout, L., Covaci, A. and Van Der Borght, M. (2021). Chemical safety of black soldier fly larvae (*Hermetia illucens*), knowledge gaps and recommendations for future research: a critical review. *Journal of Insects as Food and Feed* 7(4), 383–396.

Liu, H., Tan, B., Kong, X., Li, J., Li, G., He, L., Bai, M. and Yin, Y. (2020). Dietary insect powder protein sources improve protein utilization by regulation on intestinal amino acid-chemosensing system. *Animals* 10(9), 1590.

Liu, Q., Tomberlin, J. K., Brady, J. A., Sanford, M. R. and Yu, Z. (2008). Black soldier fly (Diptera: Stratiomyidae) larvae reduce *Escherichia coli* in dairy manure. *Environmental Entomology* 37(6), 1525–1530.

Mancuso, T., Pippinato, L. and Gasco, L. (2019). The European insects sector and its role in the provision of green proteins in feed supply. *Quality – Access to Success* 20, 374–381.

Meyer, S., Gessner, D. K., Braune, M. S., Friedhoff, T., Most, E., Höring, M., Liebisch, G., Zorn, H., Eder, K. and Ringseis, R. (2020). Comprehensive evaluation of the metabolic effects of insect meal from *Tenebrio molitor* L. in growing pigs by transcriptomics, metabolomics and lipidomics. *Journal of Animal Science and Biotechnology* 11, 20.

Meyer-Rochow, V. B. and Changkija, S. (1997). Uses of insects as human food in Papua New Guinea, Australia, and North-East India: cross-cultural considerations and cautious conclusions. *Ecology of Food and Nutrition* 36(2–4), 159–185.

Miech, P., Lindberg, J. E., Berggren, Å., Chhay, T. and Jansson, A. (2017). Apparent faecal digestibility and nitrogen retention in piglets fed whole and peeled Cambodian field cricket meal. *Journal of Insects as Food and Feed* 3(4), 279–288.

Myers, H. M., Tomberlin, J. K., Lambert, B. D. and Kattes, D. (2008). Development of black soldier fly (Diptera: Stratiomyidae) larvae fed dairy manure. *Environmental Entomology* 37(1), 11-15.

Neumann, C., Velten, S. and Liebert, F. (2018). N balance studies emphasize the superior protein quality of pig diets at high inclusion level of algae meal (*Spirulina platensis*) or insect meal (*Hermetia illucens*) when adequate amino acid supplementation is ensured. *Animals* 8(10), 1-14.

Newton, G. L., Booram, C. V., Barker, R. W. and Hale, O. M. (1977). Dried *Hermetia illucens* larvae meal as a supplement for swine. *Journal of Animal Science* 44(3), 395-400.

Newton, L., Sheppard, C., Watson, D. W., Burtle, G. and Dove, R. (2005). *Using the Black Soldier Fly, Hermetia illucens, as a Value-Added Tool for the Management of Swine Manure*. Animal and Poultry Waste Management Centre, North Carolina State University, Raleigh, NC.

OECD & Food Agriculture Organization of the United Nations, R (2022). *OECD-FAO Agricultural Outlook* (Edition 2022). OECD Agriculture Statistics (database), https://doi.org/10.1787/13d66b76-en (accessed on 28 January 2024).

Oonincx, D. G. A. B., Van Huis, A. and Van Loon, J. J. A. (2015). Nutrient utilisation by black soldier flies fed with chicken, pig, or cow manure. *Journal of Insects as Food and Feed* 1(2), 131-139.

Peng, B. Y., Su, Y., Chen, Z., Chen, J., Zhou, X., Benbow, M. E., Criddle, C. S., Wu, W. M. and Zhang, Y. (2019). Biodegradation of polystyrene by dark (*Tenebrio obscurus*) and yellow (*Tenebrio molitor*) mealworms (Coleoptera: Tenebrionidae). *Environmental Science and Technology* 53(9), 5256-5265.

Perednia, D. A. (2017). A comparison of the greenhouse gas production of black soldier fly larvae versus aerobic microbial decomposition of an organic feed material. *Research and Reviews: Journal of Ecology and Environmental Sciences* 5(3), 10-16.

Poelaert, C., Beckers, Y., Despret, X., Portetelle, D., Francis, F. and Bindelle, J. (2016). In vitro evaluation of fermentation characteristics of two types of insects as potential novel protein feeds for pigs. *Journal of Animal Science* 94(Suppl. 3), 198-201.

Ponce-Reyes, R. and Lessard, B. (2021). *Edible Insects - A Roadmap for the Strategic Growth of an Emerging Australian Industry*. CSIRO, Canberra.

Purschke, B., Scheibelberger, R., Axmann, S., Adler, A. and Jäger, H. (2017). Impact of substrate contamination with mycotoxins, heavy metals and pesticides on the growth performance and composition of black soldier fly larvae (*Hermetia illucens*) for use in the feed and food value chain. *Food Additives and Contaminants: Part A, Chemistry, Analysis, Control, Exposure and Risk Assessment* 34(8), 1410-1420.

Rahimi, P., Islam, M. S., Duarte, P. M., Tazerji, S. S., Sobur, M. A., El Zowalaty, M. E., Ashour, H. M. and Rahman, M. T. (2022). Impact of the COVID-19 pandemic on food production and animal health. *Trends in Food Science and Technology* 121, 105-113.

Rumpold, B. A. and Schlüter, O. K. (2013). Potential and challenges of insects as an innovative source for food and feed production. *Innovative Food Science and Emerging Technologies* 17, 1-11.

Scheibelberger, R., Axmann, S., Adler, A. and Jäger, H. (2017). Impact of substrate contamination with mycotoxins, heavy metals and pesticides on the growth performance and composition of black soldier fly larvae (*Hermetia illucens*) for use in the feed and food value chain. *Food Additives and Contaminants: Part A* 34, 1410-1420.

Sogari, G., Amato, M., Palmieri, R., Hadj Saadoun, J., Formici, G., Verneau, F. and Mancini, S. (2023). The future is crawling: evaluating the potential of insects for food and feed security. *Current Research in Food Science* 6, 100504.

Swiatkiewicz, S., Swiatkiewicz, M., Arczewska-Wlosek, A. and Jozefiak, D. (2015). Chitosan and its oligosaccharide derivatives (chito-oligosaccharides) as feed supplements in poultry and swine nutrition. *Journal of Animal Physiology and Animal Nutrition* 99(1), 1-12.

Tan, X., Yang, H. S., Wang, M., Yi, Z. F., Ji, F. J., Li, J. Z. and Yin, Y. L. (2020). Amino acid digestibility in housefly and black soldier fly prepupae by growing pigs. *Animal Feed Science and Technology* 263, 114446.

Van Der Fels-Klerx, H. J., Camenzuli, L., Belluco, S., Meijer, N. and Ricci, A. (2018). Food safety issues related to uses of insects for feeds and foods. *Comprehensive Reviews in Food Science and Food Safety* 17(5), 1172-1183.

Van Huis, A. (2013). Potential of insects as food and feed in assuring food security. *Annual Review of Entomology* 58, 563-583.

Van Huis, A. (2016). Edible insects are the future? *Proceedings of the Nutrition Society* 75(3), 294-305.

Van Huis, A., Van Itterbeeck, J., Klunder, H., Mertens, E., Halloran, A., Muir, G. and Vantomme, P. (2013). *Edible Insects: Future Prospects for Food and Feed Security*. FAO Forestry Paper 171. Food and Agriculture Organization of the United Nations, Rome.

Van Zanten, H. H. E., Bikker, P., Meerburg, B. G. and De Boer, I. J. M. (2018). Attributional versus consequential life cycle assessment and feed optimization: alternative protein sources in pig diets. *International Journal of Life Cycle Assessment* 23(1), 1-11.

Veldkamp, T. and Bosch, G. (2015). Insects: a protein-rich feed ingredient in pig and poultry diets. *Animal Frontiers* 5, 45-50.

Veldkamp, T., Dong, L., Paul, A. and Govers, C. (2022). Bioactive properties of insect products for monogastric animals – a review. *Journal of Insects as Food and Feed* 8(9), 1027-1040.

Veldkamp, T. and Vernooij, A. G. (2021). Use of insect products in pig diets. *Journal of Insects as Food and Feed* 7(5), 781-793.

Verbeke, W., Spranghers, T., De Clercq, P., De Smet, S., Sas, B. and Eeckhout, M. (2015). Insects in animal feed: acceptance and its determinants among farmers, agriculture sector stakeholders and citizens. *Animal Feed Science and Technology* 204, 72-87.

Weinrich, R. and Busch, G. (2021). Consumer knowledge about protein sources and consumers' openness to feeding micro-algae and insects to pigs and poultry. *Future Foods* 4, 100100.

Yu, M., Li, Z., Chen, W., Rong, T., Wang, G. and Ma, X. (2019). *Hermetia illucens* larvae as a potential dietary protein source altered the microbiota and modulated mucosal immune status in the colon of finishing pigs. *Journal of Animal Science and Biotechnology* 10, 50.

Yu, M., Li, Z., Chen, W., Rong, T., Wang, G., Wang, F. and Ma, X. (2020). Evaluation of full-fat *Hermetia illucens* larvae meal as a fishmeal replacement for weanling piglets: effects on the growth performance, apparent nutrient digestibility, blood parameters and gut morphology. *Animal Feed Science and Technology* 264, 114431.

Chapter 4

Black soldier fly meal: an alternative protein source for pigs

S. Struthers, and J. G. M. Houdijk, Scotland's Rural College (SRUC), UK; and H. N. Hall, Anpario plc, UK

1   Introduction

2   Nutritional composition of black soldier fly

3   Production performance of pigs fed diets containing black soldier fly larvae

4   Benefits of using black soldier fly

5   Challenges

6   Applications

7   Conclusion

8   Acknowledgements

9   References

1   Introduction

The rise in the global human population, in conjunction with higher standards of living, has increased the demand for animal protein for human consumption. The overall demand for protein annually is estimated to be 202 million tonnes but is expected to reach 435 million tonnes by 2050 (Boland et al., 2013; Henchion et al., 2017). It is predicted that a large proportion of this demand will be for pork and poultry products (Kim et al., 2019), given that these are the more efficient land-based, farmed animals for converting feed protein into animal protein, but this would result in increased pressure on sustainable protein sourcing of their nutrition.

Pork is currently the second most consumed protein (globally), behind poultry. It is projected to grow by 11% to 129 million tonnes by 2032 (OECD/FAO, 2023), and as a versatile and lean meat, pork can be highly efficient to produce. Currently, most of the protein requirement in pig diets is fulfilled by plant sources, predominantly soya bean meal (SBM) (Boland et al., 2013; Kim et al., 2019; Alagappan et al., 2022). However, using soya products in

http://dx.doi.org/10.19103/AS.2024.0139.05

livestock feed is largely considered unsustainable due to increasing demand for SBM without an increase in supply, which drives prices upwards, but also the significant climate impact of land use change especially for meeting this additional SBM demand, as well as the carbon footprint of the associated transport distances (Ffoulkes et al., 2021; Gupta et al., 2021). Fish meal is no longer routinely included in pig diets due to unfavourable raw material costs, as the majority of FM produced is used in aquaculture, but it is known to be a valuable and well balanced protein for young pigs especially in creep and post-weaning diets. In summary, the increased protein requirements for human consumption drives a need for alternative sustainable, protein-rich, raw materials to feed livestock.

Insects, especially as it can be argued that using pigs as bio-converters to produce high-value protein products from nutritional resources that are not suitable for direct consumption by humans, will assist in the overall sustainability of animal protein production. Since humans can readily utilise soya as a source of protein, following appropriate food preparation, more protein sources are required which do not compete with human edible sources. These alternative proteins should be suitable for feeding, cost effective and have improved sustainability credentials compared to traditional sources such as SBM.

Black soldier fly (BSF; *Hermetia illucens*) is a promising alternative protein for livestock diets. BSF can be reared on a variety of co-product streams from the worldwide food and drink industries, as well as from household and retail food surplus. This allows for bioconversion of a wide array of organic substrates, converting this material from a relatively low protein value, in terms of variable quality, into a source of much less variable, high-quality nutrients for human consumption (Spranghers et al., 2017). BSF reared on organic substrates is rich in crude protein and fats and has a nutritional profile comparable to that of SBM and FM in terms of amino acid profile. Therefore, it offers excellent soya-replacement potential (Spranghers et al., 2017; Heuel et al., 2022).

2 Nutritional composition of black soldier fly

The nutritional composition and dietary value of BSF have been studied and reviewed in depth in several papers (Barragan-Fonseca et al., 2017; Spranghers et al., 2017; Veldkamp and Vernooij, 2021; Hawkey and Hall, 2023). However, papers reviewing the commercial replacement of traditional protein sources in pig diets in a commercial setting are still relatively lacking. As protein makes up the largest component of BSF on a dry matter basis, followed by fat (Barragan-Fonseca et al., 2017), it would also require post-harvest processing to best fit the commercial diet of pigs and more closely resemble current protein sources; this would then enable the most straightforward nutritional comparison and exchange of the tradition protein source for BSF in the diet. The average crude protein and fat content in BSF are 40% and 26% on a dry matter basis, respectively; however, these values can vary based on the feed substrate and insect developmental stage at the time of harvest (Barragan-Fonseca et al., 2017; Hawkey and Hall, 2023).

BSF's amino acid profile is comparable to that of SBM and fishmeal (FM) (Hawkey and Hall, 2023). The amino acid content of BSF also does not vary much between studies; however, specific amino acid levels can differ due to feed substrate (Barragan-Fonseca et al., 2017; Spranghers et al., 2017). BSF protein contains 10 essential amino acids and has higher levels of histidine, methionine, and tryptophan when compared to SBM (Hawkey and Hall, 2023).

Saturated fatty acids comprise most of the total fat content in BSF (Barragan-Fonseca et al., 2017), with the major fatty acids being palmitic, oleic, and linoleic acids. BSF also has a high level of lauric acid, a medium-chain saturated fatty acid, which has been reported to have anti-microbial and anti-inflammatory properties (Spranghers et al., 2017).

The mineral content of BSF is generally higher than other insect species used as feed (Finke, 2013). BSF larvae, in particular, have high calcium content due to their mineralised exoskeleton (Finke, 2013); however, some of this calcium is bound to chitin, the structural carbohydrate found in insects and shellfish, making its bioavailability to the pig limited as, like most animals, it does not produce chitinases endogenously (Rathore and Gupta, 2015).

Chitin is a long-chain polysaccharide found within the exoskeleton of insects and shellfish, including prawns and lobsters (Komi et al., 2018). Chitin is resistant to degradation by the animal in the digestive tract; however, it has been shown to promote hindgut fermentation of short-chain fatty acids and have immunomodulatory effects (Komi et al., 2018; Yu et al., 2020b). This might suggest that chitin can exert prebiotic properties. Chitin has been shown to have value once extracted from BSF for use in industrial processes (bioplastics) and animal diets but at lower levels. When used to modulate gut health, it has the potential to bind endotoxins and mycotoxins.

When estimating the crude protein content of insect products, it is worthwhile noting that chitin affects this calculation due to its high non-protein nitrogen (N) content. Therefore, N should be multiplied by a factor of 5.60 for BSF protein and 4.76 for whole larvae rather than the usual 6.25 factor commonly used (Janssen et al., 2017; Ewald et al., 2020). As chitin content varies with insect life stage, it is also important to note that comparisons between sources may be affected differently. Evaluating insects on a true protein (sum of amino acids) basis over a crude protein basis is recommended to avoid these inaccuracies (Hawkey et al., 2021). This further allows better appreciation of soya replacement potential, the basis of which is digestible lysine, and not crude protein.

3 Production performance of pigs fed diets containing black soldier fly larvae

The existing literature demonstrates that BSF larvae are suitable to be included as a feed ingredient in pig diets (Newton et al., 1977; Spranghers et al., 2018; Biasato et al., 2019; Zhu et al., 2022; van Heugten et al., 2022). Research has shown that BSF's nutritional value, digestibility, and palatability (particularly larvae) are comparable to SBM and FM, highlighting its potential to partially replace these feed ingredients without adversely affecting performance and improving sustainability.

3.1 Growth performance of pigs fed diets containing protein from black soldier fly

In nursing piglets, partial replacement (3.5%) of FM in the creep diet with BSF larvae did not affect growth performance (Driemeyer, 2016) (Table 1).

This has also been observed in weaned piglets. Biasato et al. (2019) found that the inclusion of up to 10% partially defatted BSF larvae meal (replacing 60% of the SBM inclusion) in weaned piglet diets had no overall effect on growth performance. Average daily feed intake (ADFI) showed a linear response to increasing BSF larvae meal levels from 24 to 61 days post-weaning, with piglets fed 10% BSF larvae meal having the highest ADFI. This increase was attributed to increased palatability of the diet by the inclusion of partially defatted BSF. Spranghers et al. (2018) also reported no effect of full-fat (4% and 8% inclusion) and defatted (5.4% inclusion) BSF prepupae meal on the growth performance of weaned piglets when replacing whole toasted soya beans. However, the authors did note that, despite the lack of statistical differences between the treatment groups, piglets fed full-fat BSF prepupae meal at 4% or 8% inclusion gained less weight compared to control piglets fed toasted soya beans and piglets fed defatted BSF prepupae meal. Piglets fed full-fat BSF prepupae meal at 8% inclusion also had lower ADFI. They suggested this may have been caused by reduced palatability due to the large amount of free medium-chain fatty acids present in the feed containing full-fat BSF prepupae (Dierick et al., 2002; Spranghers et al., 2018). However, Newton et al. (1977) reported no differences in palatability between piglets fed BSF prepupae or SBM diets.

More recently, Boontiam et al. (2022) studied the effect of supplementing 6% or 12% full-fat BSF larvae on the growth performance of weaned piglets under poor hygiene conditions. Compared to the negative control (poor hygiene and 0% BSF larvae), pigs fed BSF larvae had higher body weights and ADFI from 1 to 28 days post-weaning. Dietary inclusion of BSF also improved body weight and ADFI from 15 to 28 days post-weaning compared to the positive control group (good hygiene and 0% BSF larvae). Over the entire experimental period (1 to 28 days post-weaning), no differences in growth

Table 1 Performance of pigs fed protein from black soldier fly (BSF).

Pig growth stage	Age at weaning	Experimental period	BSF life stage	BSF feed form	BSF inclusion (%)	Target replacement	Replacement (%)	Results	Reference
Nursing		10–28 d of age	Larvae	Meal	0, 3.5	Fishmeal	3.5	Growth performance is not affected. Haematological and biochemical parameters are not affected.	Driemeyer (2016)
Weaned piglets	20 d	1–61 d post-wean	Larvae	Partially defatted meal	0, 5, 10	Soybean meal	0, 31, 62	Growth performance is not affected. Haematological parameters are not affected. Gut morphology and histology are not affected. Trend for higher FI with 10% BSF inclusion.	Biasato et al. (2019)
Weaned piglets	28 d	1–28 d post-wean	Larvae	Full-fat meal	0, 6, 12	Soybean meal	0, 20, 40	Growth performance is not affected. Diarrhoea rate was reduced in pigs fed 6 or 12% BSF.	Boontiam et al. (2022)
	21 d	0–42 d post-wean	Larvae	Full-fat meal	0.5–15 (depending on phase)	Dried whey, fishmeal, blood meal, blood plasma	0, 25, 50	Growth performance is not affected.	Crosbie et al. (2021)
	32 d	0–27 d post-wean	Larvae	Meal	0, 5, 10, 20	Fishmeal	0, 25, 50, 100	Growth performance is not affected except for reduced ADG with 5% BSF inclusion.	Håkenåsen et al. (2021)
	28 d	1–28 d post-wean	Larvae	Meal	0, 4.5, 9.1, 13.7, 18.3	Soybean meal	0, 25, 50, 75, 100	ADG and ADFI were improved with 4.5 and 9.1% BSF inclusion. FCR was not affected.	Liu et al. (2023)

(Continued)

Table 1 (*Continued*)

Pig growth stage	Age at weaning	Experimental period	BSF life stage	BSF feed form	BSF inclusion (%)	Target replacement	Replacement (%)	Results	Reference
	21 d	1–15 d post-wean	Prepupae	Full-fat meal	0, 4, 8	Toasted soybean	0, 50, 100 72	Growth performance is not affected. Gut morphology and histology are not affected.	Spranghers et al. (2018)
	21 d	1–15 d post-wean	Prepupae	Defatted meal	5.4	Toasted soybean	0, 50, 100 72	Growth performance is not affected. Gut morphology and histology are not affected.	Spranghers et al. (2018)
	21 d	1–28 d post-wean	Larvae	Full-fat meal	0 1, 2, 4	Fishmeal	0, 25, 50, 100	Growth performance is not affected. Serum total protein and globulin levels were highest while urea and triglyceride levels were lowest at 2% BSF inclusion. Jejunal villi height was the longest with 2% BSF inclusion. Inclusion of 2% BSF affected specific gut microbial populations, metabolic profiles, and mucosal immune gene expression.	(Yu et al. (2020a,b)

Pig growth stage	Age at weaning	Experimental period	BSF life stage	BSF feed form	BSF inclusion (%)	Target replacement	Replacement (%)	Results	Reference
Grower		9 wk	Larvae	Full-fat meal	0, 9, 12, 14.5, 18.5	Fishmeal	0, 25, 50, 75, 100	Growth performance is not affected. FCR was higher at 14.5 and 18.5% inclusion. Neutrophil count was higher and platelet counts were lower at 14.5 and 18.5% BSF inclusion. No other haematological parameters were affected.	Chia et al. (2019)
Grower		4 wk	Larvae		0, 3	Poultry offal	100	Growth and intake were not affected though FCR was increased.	Go et al. (2022)
Finisher		46 d	Larvae	Dried powder meal	0, 4, 8	Soybean meal	0, 18, 36	Final BW and ADG were the highest and FCR lowest at 4% inclusion. BSF inclusion positively impacted meat quality. BSF inclusion impacted microbial populations, metabolic profiles, and immune gene expression in the gut.	Yu et al. (2019a, 2019b)
Finisher		14 wk	Larvae	Full-fat meal	0, 6, 9, 12, 14	Fishmeal	0, 25, 50, 75, 100	ADFI was not affected; the Inclusion of 9 to 14% BSF improved ADG and reduced FCR.	Chia et al. (2021)

ADG = average daily gain; ADFI = average daily feed intake; BW = body weight; FCR = feed conversion ratio (feed over gain); FI = feed intake performance were found between the treatment groups; however, diarrhoea rate was significantly reduced in piglets fed BSF larvae (Boontiam et al., 2022). This suggests that feeding BSF can help improve piglet health and mitigate the effects of housing pigs in poor hygiene conditions (lighter body weight, increased diarrhoea, decreased nutrient utilisation). This is likely due to the antimicrobial properties of the fatty acid profile in the full-fat BSF, though prebiotic benefits arising from chitin cannot be excluded.

Furthermore, the inclusion of full-fat BSF larvae meal to partially replace (25% or 50%) animal protein sources (FM, spray-dried blood meal, and blood plasma) in nursery pig diets supported growth performance suggesting that dietary animal protein sources could be replaced by as much as 50% with full-fat BSF larvae meal (inclusion of up to approximately 15% in the diet) without negative consequences (Crosbie et al., 2021).

Partially replacing FM with 2% full-fat BSF larvae meal positively impacted ADG and FCR (F:G) in weaned piglets during the first two weeks post-weaning, with ADG being highest and FCR lowest in piglets fed 2% BSF larvae meal (Yu et al., 2020a). Over the 28-day post-weaning experimental period, no differences were found between the treatment groups fed 0% to 4% full-fat BSF larvae meal in replacement of 0% to 100% of the FM inclusion (Yu et al., 2020a). This is similar to what was reported by Håkenåsen et al. (2021) except the BSF inclusion rate was higher (0%, 5%, 10%, 20% of the diet), the BSF still replaced FM at 0%, 25%, 50%, 75%, and 100%. Interestingly, the only effect was seen for ADG over the 27-day post-weaning experimental period with piglets fed 5% full-fat BSF larvae meal having lower ADG compared to the control diet with the 10% and 20% BSF inclusion groups being intermediate (Håkenåsen et al., 2021). As this is the lowest BSF inclusion studied in this chapter, it may be that a higher inclusion and replacement of traditional protein sources are needed for the accurate evaluation of novel proteins in the diet.

Liu et al. (2023) investigated the effect of total replacement of SBM with increasing levels of BSF larvae meal (0% to 18.3%). The authors found that ADG and ADFI improved in piglets fed 4.5 and 9.1% BSF larvae meal (replacing 25% and 50% SBM, respectively), although FCR was unaffected.

In grower pigs, replacing up to 100% of fishmeal in the diet with full-fat BSF larvae meal (18.5% inclusion) did not negatively impact growth performance (Chia et al., 2019). Pigs fed 9% BSF larvae meal had better FCR than those fed 14.5 or 18.5% BSF larvae meal; however, all three groups were not different from the fishmeal control diet (Chia et al., 2019). Haematological parameters were generally unaffected, except for neutrophil and platelet counts, which were higher in the 14.5 and 18.5% BSF larvae meal inclusion groups (representing 75% and 100% fishmeal replacement, respectively).

Go et al. (2022) found that ADG was reduced during the first 2 weeks of feeding 3% BSF larvae compared to pigs fed a control diet containing poultry offal (100% replacement). However, by the end of the 4-week experimental period, no differences in ADG were noted. Pigs fed the 3% BSF larvae diet had lower FCR (G:F) than controls over the 4-week experiment. These minor differences in growth performance were attributed to anti-nutritional factors (ANFs) in the BSF, such as chitin. Chitin is a component of the insect's exoskeleton, and if not digested, it can bind to proteins and reduce digestibility (Wang and Shelomi, 2017).

In finisher pigs, partially replacing soybean meal with 4% dried BSF larvae meal (18% replacement) improved body weight and ADG and reduced FCR (F:G) compared to the control and 8% BSF inclusion diets (Yu et al., 2019a). This agrees with what has been reported in more recent studies. Chia et al. (2021) found replacing up to 100% fishmeal with up to 14% full-fat BSF larvae meal did not affect body weight or ADFI; however, the higher levels of BSF inclusion (9% to 14%) improved ADG and lowered FCR (F:G). Yu et al. (2019a) suggested that the differences between their treatment groups could be related to the upregulated expression of genes related to lipogenic potential and muscle fibre composition. They also attributed the lack of effect of the 8% BSF inclusion to higher chitin levels in the BSF diet. The improved growth performance in pigs fed increasing levels of BSF larvae meal in the study by Chia et al. (2021) was attributed to increased palatability.

The aforementioned studies were all carried out with the idea of using BSF as an alternative protein source. One study though looked at the use of BSF oil. Thus, when BSF larvae were included in the diet of weaned piglets as an extracted insect oil to replace corn oil, body weight, and ADG linearly increased with increasing BSF oil inclusion (0%, 2%, 4%, 6%). Serum cholesterol and platelet count also increased linearly with increasing BSF oil inclusion (van Heugten et al., 2022). This suggests that BSF larvae oil is at least as palatable as corn oil suitable to be considered as an alternative energy source.

Overall, these studies suggest that the inclusion of BSF in the diets of piglets both immediately post-weaning and at later ages does not adversely affect growth performance. The number of studies carried out in growing and finisher pigs is rather limited and so more work would be required to confirm the benefits of replacing SBM with BSF in these diets.

3.2 Nutrient digestibility

From the 12 studies reviewed, that reported pig performance when replacing traditional protein sources (one study included reviewed oil replacement), only three of these also reported nutrient digestibility. Commercial scale studies are still required to further evaluate the nutrient digestibility of BSF inclusion in pig diets, however we have summarised the current findings below.

Nutrient digestibility was not influenced by the inclusion of partially defatted 10% BSF larvae meal in the diet of weaned piglets (Biasato et al., 2019).

Dietary supplementation of 6% or 12% full-fat BSF larvae meal improved dry matter, crude protein, and ether extract digestibility in weaned piglets when compared to the negative control group (poor hygiene and 0% BSF); however, the BSF inclusion groups did not differ from the positive control (good hygiene and 0% BSF) (Boontiam et al., 2022). This improved digestibility is likely due to the improved gut health of piglets fed full-fat BSF larvae, which would help reduce unwanted inflammation and can result in increased energy partitioning to growth.

Including 5.4% defatted BSF larvae meal in weaned piglet diets resulted in equal or improved ileal nutrient digestibility compared to the control diet containing toasted soybean (Spranghers et al., 2018). Including full-fat BSF larvae meal at 4% or 8% reduced ileal energy digestibility; however, 4% BSF larvae meal inclusion increased ileal crude protein digestibility (Spranghers et al., 2018).

The dietary BSF inclusion in these studies is relatively low. At higher levels of BSF larvae inclusion (33%) in a study from almost 50 years ago, Newton et al. (1977) found reduced dry matter, increased ether extract, and similar crude protein digestibility in weaned piglets compared to those fed soybean meal. These differences may be due to changes in gut microbiome following the feeding of a novel feed ingredient.

3.3 Gut health, haematological parameters, and meat quality

Driemeyer (2016) found that including 3.5% BSF larvae meal in nursing piglet diets (replacing 100% fishmeal inclusion) did not affect haematological or biochemical parameters. The authors did observe increasing haemoglobin and haematocrit levels over the experimental period with BSF inclusion. Although the differences were not statistically significant, the authors suggested that they may be biologically important as higher haemoglobin and haematocrit levels can be indicators of immunological stress; however, the authors also noted that none of the animals showed physical signs of distress (Driemeyer, 2016).

Biasato et al. (2019) found that BSF inclusion in the diet of weaned piglets did not affect haematological parameters except for monocytes and neutrophils, which showed linear and quadratic responses to increasing BSF inclusion, respectively. Regardless, all of the haematological parameters recorded fell within the normal physiological range for pigs, suggesting that BSF inclusion did not negatively impact the piglet's health (Biasato et al., 2019). Inclusion of up to 10% partially defatted BSF larvae meal also did not influence gut morphology or histology (Biasato et al., 2019).

Serum cholesterol linearly increased in weaned piglets fed increasing levels of BSF larvae oil (van Heugten et al., 2022). Platelet count also tended to increase linearly with increasing BSF oil inclusion. BSF larvae oil inclusion affected no other haematological or serum parameters (van Heugten et al., 2022).

Yu et al. (2020a, 2020b) reported that weaned piglets fed 2% full-fat BSF larvae meal had higher serum total protein and globulin levels than piglets fed FM. Gut morphology was also influenced, with piglets fed 2% BSF larvae meal having longer jejunal villi heights than piglets fed fishmeal or 1% BSF larvae meal (Yu et al., 2020a). Including 2% BSF larvae meal also affected specific microbial populations, their metabolic profiles, and mucosal immune gene expressions in the gut (Yu et al., 2020b).

In grower pigs, neutrophil counts were increased, and platelet counts were reduced in pigs fed 14.5 or 18.5% full-fat BSF larvae meal instead of fishmeal (Chia et al., 2019), which coincided with greater (more detrimental) FCR. Since no other haematological were affected by BSF inclusion in this study, an elevated neutrophil level might be an indication of a stress response (Widowski et al, 1989), in accord with as suggested by Driemeyer (2016).

In finisher pigs, Yu et al. (2019a) found that including dried BSF larvae meal (4% or 8%) positively influenced carcass traits and meat quality as loin eye area and marbling scores were higher with BSF inclusion. Similar to what was found in the weaned piglets, Yu et al. (2019b) reported that BSF inclusion in finisher pig diets altered the microbial populations, their metabolic profiles, and mucosal immune gene expression in the gut.

4 Benefits of using black soldier fly

4.1 Reducing the use of traditional protein sources, soya bean meal, and fishmeal

This chapter evaluates peer-reviewed papers where BSF has been utilised to replace mainly SBM or FM in pig diets and these clearly indicate that while there is evidence to support the use of BSF as a suitable alternative, there is still a significant degree of variability regarding the upper limit and dose-dependency of this approach. For most studies, the largest inclusion of BSF used, ranging from 5% to 20%, did not impact growth performance. However, in some studies, positive effects on performance were observed at levels below the maximum tested. This supports that, as is the case with any ingredient, there are constraints to consider, which in the case of BSF are likely coming from the level of post-harvest processing regarding protein and oil extraction and chitin levels, which can be a result of variation in growth stage and processing. As such, owing to its positive effect on protein concentration, defatting BSF would be expected to increase its suitability in the commercial pig diet and therefore its soya replacement potential, though it should be noted that in the inclusion levels of full-fat, partially defatted, or defatted BSF reported to date, an upper limit on SBM replacement in young pigs seems not to have been demonstrated yet. There are also trade-offs to be considered between benefits and constraints from both BSF and the replaced ingredients. For the BSF, these trade-offs may come from the level of chitin, which at low inclusion level might be providing prebiotic benefits but over a threshold might result in fibre-based constraints. As such, replacing SBM with BSF will inevitably reduce the levels of fibre in the diet. Furthermore, if the replaced ingredient is arguably of greater digestibility, replacement on a digestible amino acid basis would be expected to detriment feed efficiency, as observed for BSF inclusion at the expense of poultry offal (Go et al., 2022). Finally, it should also be noted that whilst in some studies all the SBM was safely replaced, i.e. without impacting performance, this will be sensitive to the basal level of SBM in the first place. The lower the latter, the more likely it can be completely replaced with an alternative.

The vast majority of growth performance studies using BSF have been on peri-weaning pigs, whilst only a few have been carried out on grower and finisher pigs. Whilst arguably the level of SBM in pig rations reduces as pigs move through the different feeding phases, the greater feed intake of grower and particularly finisher pigs compared to peri-weaning pigs means that most SBM is being used in the grower and finisher phase. Only one study was identified in finisher pigs, where up to 8% of BSF meal was replacing up to 36% of SBM, with improved FCR and ADG at intermediate BSF levels (Yu et al., 2019). More studies in grower and finisher pigs are required to identify if a greater replacement of SBM for BSF results in similar performance, as this single study might suggest it may not be the case. The notion that alternative feedstuffs each have their own constraints when being used as 'soya replacers' comes largely from studies where these alternatives are addressed in isolation. However, an alternative approach would be to set a limit in the diet formulation to remove or restrain SBM inclusion and provide all other alternatives. A best-cost formulation may also look to include sustainability metrics to create a 'low carbon' or 'home-grown' diet which has other benefits to sustainable livestock production. Including BSF and other alternatives would help to result in a greater reduction in SBM, with the emphasis on using upper limits of each alternative below their anti-nutritional thresholds but so as to not produce a diet with too many ingredients. Such approaches have been shown successful in replacing all SBM in fast growing broilers (Houdijk et al., 2024), which arguably have a digestive system similar to that of pigs. Thus, realising that this can be achieved in broilers, this opens the notion to combine BSF with other protein sources such as pulses and oilseed meals from non-soya origin (e.g. rapeseed meal, sunflower meal). Indeed, within a background of a combination of SBM and rapeseed meal as main protein sources, a complete replacement of SBM with home-grown pulses for growing and finishing pigs has been observed under both experimental conditions (Smith et al., 2013; White et al., 2015) and commercial conditions (Houdijk et al., 2013).

Like SBM, FM has also been under pressure from a sustainability perspective, given that fish stocks are a finite resource (concerns of overfishing). FM also has environmental implications associated with its transport and processing (especially drying), and such factors have made it increasingly expensive and supply volatile. As such, we usually only find FM in the most specialist rations for young pigs (creep feeds, weaner feeds) as most of the FM available in the market is used in pet nutrition and aquaculture diets. The historic use of FM in young pig rations was not only driven by its high biological value, but also its associated health benefits, including gut health (see below) arising from a favourable fatty acid and amino acid profile. Therefore, replacing FM with BSF meal could not only provide a cost-effective alternative as well as one with improved sustainability credentials and supply benefits.

4.2 Gut health

Several studies have reported on the gut health benefits of BSF and extracted products such as insect oil, chitin, and antimicrobial peptides (AMPs) (Gasco et al., 2018; Biasato et al., 2019; Boontiam et al., 2022). While further work is needed to understand whether there is greater value in feeding these as extracted components or as the whole insect it should be noted that regulatory hurdles may also play a role in this decision.

As mentioned, chitin is the main carbohydrate incorporated in the insect exoskeleton and can impact protein and mineral bioavailability (Finke, 2007; Jonas-Levi and Martinez, 2017; Henriques et al., 2020). While pigs do not sufficiently produce chitinases to easily digest the chitin, it has been shown that the gut microbiota can secrete chitinase and gain value in this component (Šimůnek et al., 2001). The extent of this is likely related to the age of the pig and gut health before feeding a chitin-based compound.

Antimicrobial peptides are known to be produced by insects (Harlystiarini et al., 2019) as part of their immune response. These proteins inhibit the growth of harmful pathogens in the insect (Lu et al., 2014), but they also confer benefits to livestock feeding and may be a factor in the improvements in health seen in the papers mentioned above, most notably (Boontiam et al., 2022). *In vitro* BSF extracts have been shown to be strongly antibacterial against species of *Salmonella* and *E. coli* (Harlystiarini et al., 2019); both pathogens are important in pig production and may provide added value when feeding whole or extracted BSF products.

The composition of BSF oil is similar to that of coconut oil (Dayrit, 2015), predominantly containing lauric acid, with up to 52% recently having been reported (Ewald et al., 2020). Lauric acid is known to have antimicrobial properties (Spranghers et al., 2018). The content of lauric acid in BSF fat is affected by diet type and suitability but it is positively correlated to larvae weight and is hypothesised to be synthesised by the larvae (Spranghers et al., 2017) as it is found in high concentrations even when fed at very low levels in the insect substrate. Coconut oil and lauric acid are known to be beneficial in pig diets historically, with coconut oil being included in creep and nursery diets for antimicrobial benefits, but care must be taken to manage the antioxidant capacity of the diet when including highly saturated fatty acids.

Feeding of BSF has been shown to modify the gut microbiota (Håkenåsen et al., 2021; Boontiam et al., 2022; Liu et al., 2023), with increases seen in *Ruminococcaceae*, *Faecalibacterium*, *Butyricoccus*, and *Lactobacillus* spp. These changes may be due to the components mentioned above and changes to the macronutrients available in the diet. Further work is needed to consistently use BSF to improve gut health, but it is clear to see that gut health benefits can be gained from their feeding, and this should be harnessed to fully realise the value of BSF in pig diets.

4.3 Reducing carbon footprint of diets

It has been suggested that insect-based diets can reduce the carbon emissions of livestock feeding when replacing FM or SBM (van Huis and Oonincx, 2017; Ffoulkes et al., 2021). However, insect feedstocks and climate control in insect rearing must be considered (Oonincx, 2021) and the processing of insects post-harvest when compared to mass-produced feed protein such as SBM. In terms of locality and transport emissions, BSF can provide a highly localised protein source and may even be able to be produced on the site where the protein is to be used. This form of circular economy can make great use of locally available co-products, which would still fit with regulatory requirements within the EU. Insects such as BSF are classed as farmed livestock within the EU and as such must be reared and fed in accordance with relevant regulations (Alagappan et al., 2022). However, insects can be succesfully reared on vegetable surplus from supermarkets or pre-consumer sources. This would increase the nutrient suitability of these substrates and reduce unwanted variation in nutrient provision for livestock feeding (van Heugten et al., 2022). It may also be that insects play a role alongside anaerobic digesters, utilising waste heat and being used as a primary bio-converter in this process. We are increasingly seeing sustainable livestock farming look to utilise all on-farm primary and secondary resources to reduce waste, improve carbon capture and enhance overall resource efficiency; in this context, insects such as BSF can play a vital piece in this complicated puzzle (van Huis and Oonincx, 2017).

4.4 Animal welfare following feeding of BSF

A wealth of data supports the notion that insects support good animal welfare, especially regarding pig, poultry, and aquaculture feeding. This is mainly because most of the diet of juvenile monogastrics and fish would be supported by insects, either terrestrial or aquatic. Therefore, insect feeding supports key welfare criteria in livestock farming, providing towards freedoms 1 (freedom from thirst, hunger, and malnutrition) and 5 (freedom to express normal behaviour) (Mellor, 2016). The highest value regarding animal welfare from feeding insects may be attributed to live or whole insect feeding as they have then been attributed to providing stimulation in the form of manipulable materials in pig and poultry rearing (Star et al., 2020; Ipema et al., 2021a,b; 2022). However, wider use may be found in the use of insects in pelleted feeds, which would likely be in the form of a protein powder or extracted oil. This use may also benefit animal welfare, especially if the gut and animal health benefits reported above can be fully realised.

5   Challenges

5.1 Feed safety

A few studies have examined aspects of the safety of feeding BSF to pigs. Potential feed safety concerns such as microbial contamination and heavy metal accumulation need to be investigated and addressed (reviewed by van der Fels-Klerx et al. (2018)). An advantage is that, unlike other insect species, BSF are not considered disease vectors as the adults do not consume decayed organic material nor lay their eggs on organic material (van Huis, 2013). However, as with other animal-derived proteins, there is a risk of prion diseases, although there is (currently) no evidence that insects carry prions (DiGiacomo and Leury, 2019). BSF larvae have been found to accumulate some heavy metals such as cadmium from their diet but not others such as chromium, arsenic, nickel, and mercury (Charlton et al., 2015; Cai et al., 2018). Zinc concentration decreases in BSF larvae as its concentration increases in the rearing substrate (Diener et al., 2015). There are differing results regarding lead accumulation with some studies reporting higher concentrations in the BSF larvae than in the rearing substrate (Gao et al., 2017; Purschke et al., 2017; Cai et al., 2018) while others report the opposite (Diener et al., 2015). On a positive note regarding feed safety, BSF larvae have the ability to remove mycotoxins from contaminated feed without subsequent accumulation (Cai et al., 2018; van der Fels-Klerx et al., 2018). Additionally, BSF larvae reared on pesticide-spiked substrates did not show detectable levels of the pesticides in their tissues (Purschke et al., 2017). Thus, although when acting as bio-converters, some degree of bioaccumulation is always expected, the evidence that this is greatly impacting BSF quality is not strong. Therefore, existing quality assurance schemes in animal feed production that guarantee traditional feed safety are likely also applicable for the safe use of BSF.

5.2 Feed suitability

The knowledge of the nutritional value of BSF and any potential anti-nutritional factors it may contain will greatly determine its feed suitability. Some possible impacts of fibrous chitin have already been mentioned, including its prebiotic potential at relatively low levels and its fibre-like constraints at relatively high levels. The microbiome can make use of chitin as a source of energy by action of their chitinases. This opens the suggestions to develop chitinase as a possible feed additive, in the same way how, e.g. commercial phytases, xylanases, and proteases have been developed. This might assist the production of oligosaccharide-type structures from chitin to promote prebiotic properties and overall fermentability, and this contribution to host energy supply and physicochemical modification of gut content. Such an approach might also address the possible issue of allergen risk or allergy-promoting molecules present within the insects. Chitin may play a role in allergenicity as it is recognised by the immune system and can activate various immune cells (Komi et al., 2018). Despite chitin's role in immune and potentially allergic responses, the mechanisms by which it does so are still not fully understood (van Huis and Oonincx, 2017; van der Fels-Klerx et al., 2018). Another aspect of feed suitability is palatability. However, if palatability was a serious constraint, this would have been observed as reduction in feed intake in the performance studies undertaken. Since the latter has not been reported, it might be concluded that the palatability of BSF is unlikely a constraint.

6 Applications

6.1 Inclusion into pelleted feed

As has already been discussed earlier in the chapter, oil extracted BSF protein is a more suitable comparison to conventional protein sources (SBM and FM) due to the low level of oil in these materials. Pelleted monogastric feeds generally contain no more than 3.5% crude fat as higher levels can negatively affect pelleting and can reduce pellet quality as well as increasing the risk of oxidation and potentially leading to the early breakdown of feed vitamins.

Commercially, BSF is available and increasingly as a fat extracted protein over the whole unprocessed product. In terms of legislation, the EU classifies whole unprocessed terrestrial invertebrates and live-fed terrestrial invertebrates as currently permitted for feeding to all livestock. However, oil extraction would categorise the insect protein as processed animal protein (PAP) and is then regulated under different criteria. It is therefore in some cases more accessible to feed BSF as live and, or unprocessed rather than through the PAP regulations. For widespread use of BSF in pig diets it is likely that further processing will be needed to reduce limiting nutrients such as chitin and enable easy inclusion into pelleted feeds which makes up the majority of pig feed volume.

6.2 Live fed on farm

As noted above, this is generally an easier application due to the avoidance of the PAP regulations which came into force to reduce the risk of bovine spongiform encephalitis (BSE). Live feeding of BSF on farms is one of the main ways BSF are currently fed to livestock in the EU and UK, such as laying hens in the aim of reducing carbon footprint per egg. However, it is noted that current schemes are not fully effective due to the high costs of heating insect-rearing units and the limitations in legislation surrounding the feeding of insects. For the full value of BSF feeding to be realised it is likely that legislation surrounding permitted insect feeding substrates needs to be reviewed and relaxed to enable a wider array of lower value organic streams to be utilised.

7 Conclusion

The current body of research demonstrates that BSF has excellent potential as an insect species to be used for feeding commercial livestock, especially pigs. BSF is able to partially, and in some cases fully, replace SBM and FM without negative consequences for growth performance, gut health, or meat quality. BSF also has the ability to have less of an ecological footprint compared to the current plant protein sources but is hampered by current legislative frameworks. Ultimately, more research is needed at a commercial scale to ensure effective inclusion in commercial pig diets as a safe, cost-effective, and sustainable protein.

8 Acknowledgements

SRUC receives support from the Scottish Government (RESAS), including through an initiative around insect products for food production.

9 References

Alagappan, S., Rowland, D., Barwell, R., Mantilla, S. M. O., Mikkelsen, D., James, P., Yarger, O., and Hoffman, L. C. 2022. Legislative landscape of black soldier fly (Hermetia illucens) as feed. *J Insects Food Feed* 8:343–355.

Barragan-Fonseca, K. B., Dicke, M. and van Loon, J. J. A. 2017. Nutritional value of the black soldier fly (Hermetia illucens L.) and its suitability as animal feed: a review. *J Insects Food Feed* 3:105–120.

Biasato, I., Renna, M., Gai, F., Dabbou, S., Meneguz, M., Perona, G., Martinez, S., Lajusticia, A. C. B., Bergagna, S., Sardi, L., Capucchio, M. T., Bressan, E., Dama, A., Schiavone, A. and Gasco, L. 2019. Partially defatted black soldier fly larva meal inclusion in piglet diets: effects on the growth performance, nutrient digestibility, blood profile, gut morphology and histological features. *J Anim Sci Biotechnol* 10.

Boland, M. J., Rae, A. N., Vereijken, J. M., Meuwissen, M. P. M., Fischer, A. R. H., van Boekel, M. A. J. S., Rutherfurd, S. M., Gruppen, H., Moughan, P. J. and Hendriks, W. H. 2013. The future supply of animal-derived protein for human consumption. *Trends Food Sci Technol* 29:62–73.

Boontiam, W., Phaengphairee, P., Hong, J. and Kim, Y. Y. 2022. Full-fatted Hermetia illucens larva as a protein alternative: effects on weaning pig growth performance, gut health, and antioxidant status under poor sanitary conditions. *J Appl Anim Res* 50:732–739.

Cai, M., Hu, R., Zhang, K., Ma, S., Zheng, L., Yu, Z. and Zhang, J. 2018. Resistance of black soldier fly (Diptera: Stratiomyidae) larvae to combined heavy metals and potential application in municipal sewage sludge treatment. *Environ Sci Pollut Res* 25:1559–1567.

Charlton, A. J., Dickinson, M., Wakefield, M. E., Fitches, E., Kenis, M., Han, R., Zhu, F., Kone, N., Grant, N., Devic, E., Bruggeman, G., Prior, R. and Smith, R. 2015. Exploring the chemical safety of fly larvae as a source of protein for animal feed. *J Insects Food Feed* 1:7–16.

Chia, S. Y., Tanga, C. M., Osuga, I. M., Alaru, A. O., Mwangi, D. M., Githinji, M., Dubois, T., Ekesi, S., van Loon, J. J. A. and Dicke, M. 2021. Black soldier fly larval meal in feed enhances growth performance, carcass yield and meat quality of finishing pigs. *J Insects Food Feed* 7:433–447.

Chia, S. Y., Tanga, C. M., Osuga, I. M., Alaru, A. O., Mwangi, D. M., Githinji, M., Subramanian, S., Fiaboe, K. K. M., Ekesi, S., van Loon, J. J. A. and Dicke, M. 2019. Effect of dietary replacement of fishmeal by insect meal on growth performance, blood profiles and economics of growing pigs in Kenya. *Animals* 9. https://doi.org/10.3390/ani9100705

Crosbie, M., Zhu, C., Karrow, N. A. and Huber, L. A. 2021. The effects of partially replacing animal protein sources with full fat black soldier fly larvae meal (Hermetia illucens) in nursery diets on growth performance, gut morphology, and immune response of pigs. *Transl Anim Sci* 5. https://doi.org/10.1093/tas/txab057

Dayrit, F. M. 2015. The properties of lauric acid and their significance in coconut oil. *J Am Oil Chem Soc* 92:1–15.

Diener, S., Zurbrügg, C. and Tockner, K. 2015. Bioaccumulation of heavy metals in the black soldier fly, Hermetia illucens and effects on its life cycle. *J Insects Food Feed* 1:261–270.

Dierick, N. A., Decuypere, J. A., Molly, K., Van Beek, E. and Vanderbeke, E. 2002. The combined use of triacylglycerols containing medium-chain fatty acids (MCFAs) and exogenous lipolytic enzymes as an alternative for nutritional antibiotics in piglet nutrition. *Livest Prod Sci* 75:129–142.

DiGiacomo, K. and Leury, B. J. 2019. Review: insect meal: a future source of protein feed for pigs? *Animal* 13:3022–3030.

Driemeyer, H. 2016. Evaluation of black soldier fly (Hermetia illucens) larvae as an alternative protein source in pig creep diets in relation to production, blood and manure microbiology parameters. Available at https://scholar.sun.ac.za

Ewald, N., Vidakovic, A., Langeland, M., Kiessling, A., Sampels, S. and Lalander, C. 2020. Fatty acid composition of black soldier fly larvae (Hermetia illucens) – Possibilities and limitations for modification through diet. *Waste Manag* 102:40–47. https://doi .org/10.1016/j.wasman.2019.10.014

Ffoulkes, C., Illman, H., O'connor, R., Lemon, F., Behrendt, K., Wynn, S., Wright, P., Godber, O., Ramsden, M., Adams, J., Metcalfe, P., Walker, L.,. Gittins, J., Wickland, K., Nanua, S. and Sharples, B. 2021. Development of a roadmap to scale up insect protein production in the UK for use in animal feed. https://www.wwf.org.uk/sites/default/ files/2021-06/the_future_of_feed_technical_report.pdf

Finke, M. D. 2007. Estimate of chitin in raw whole insects. *Zoo Biol* 26:105–115.

Finke, M. D. 2013. Complete nutrient content of four species of feeder insects. *Zoo Biol* 32:27–36.

Gao, Q., Wang, X., Wang, W., Lei, C., and Zhu, F. 2017. Influences of chromium and cadmium on the development of black soldier fly larvae. *Environ Sci Pollut Res* 24:8637–8644.

Gasco, L., Finke, M. and van Huis, A. 2018. Can diets containing insects promote animal health? *J Insects Food Feed* 4:1–4.

Go, Y. B., Lee, J. H., Lee, B. K., Oh, H. J., Kim, Y. J., An, J. W., Chang, S. Y., Song, D. C., Cho, H. A., Park, H. R., Chun, J. Y. and Cho, J. H. 2022. Effect of insect protein and protease on growth performance, blood profiles, fecal microflora and gas emission in growing pig. *J Anim Sci Technol* 64:1036–1076.

Gupta, M., Hart, P., Krebsbach, S. G., Da Gama, L., Baungaard, C., Trewern, J., Halevy, S., Walsh, L., Blandon, A., Weir, C., Ryan, A., Wakefield, S., Tesco, D. E., Webb, L., Nordmann, H. D., Weis, B., Umeasiegbu, K., Ffoulkes, C., Illman, H., Behrendt, K., Godber, O., Ramsden, M., Adams, J., Metcalfe, P., Walker, L., Gittins, J., Wynn, S., O'connor, R., Lemon, F., Wickland, K., Nanua, S., Sharples, B., Wright, P., Keller, E., Perkins, R. and Salter, D. 2021. The future of feed: a WWF roadmap to accelerating insect protein in UK feeds. www.wwf.org.uk/press-release/insects-animal-feed -report. Text © 2021 WWF-UK. All rights reserved.

Håkenåsen, I. M., Grepperud, G. H., Hansen, J. Ø., Øverland, M., Ånestad, R. M. and Mydland, L. T. 2021. Full-fat insect meal in pelleted diets for weaned piglets: effects on growth performance, nutrient digestibility, gastrointestinal function, and microbiota. *Anim Feed Sci Technol* 281. https://doi.org/10.1016/j.anifeedsci.2021 .115086

Harlystiarini, H., Mutia, R., Wibawan, I. W. T. and Astuti, D. A. 2019. In vitro antibacterial activity of black soldier fly (Hermetia illucens) larva extracts against Gram-negative bacteria. *Buletin Peternakan* 43:125–129.

Hawkey, K., Brameld, J., Parr, T., Salter, A. and Hall, H. 2021. *Suitability of insects for animal feeding.Pages 26–38 in Insects as animal feed: novel ingredients for use in pet, aquaculture and livestock diets.* CABI, Wallingford, UK.

Hawkey, K. and Hall, H. 2023. Insects as animal feed. Pages 372–373 in *The Encyclopedia of Animal Nutrition.* Phillips, C., Ed. 2nd ed. CABI, GB.

Henchion, M., Hayes, M., Mullen, A. M., Fenelon, M. and Tiwari, B. 2017. Future protein supply and demand: strategies and factors influencing a sustainable equilibrium. *Foods* 6:1–21.

Henriques, B. S., Garcia, E. S., Azambuja, P. and Genta, F. A. 2020. Determination of chitin content in insects: an alternate method based on calcofluor staining. *Front Physiol* 11:1–10.

Heuel, M., Sandrock, C., Leiber, F., Mathys, A., Gold, M., Zurbrüegg, C., Gangnat, I. D. M., Kreuzer, M. and Terranova, M. 2022. Black soldier fly larvae meal and fat as a replacement for soybeans in organic broiler diets: effects on performance, body N retention, carcase and meat quality. *Br Poult Sci* 63:650–661.

Houdijk, J. G. M., Marchal, L., Bello, A., Gibbs, K. and Dersjant-Li, Y. 2024. Complete replacement of soya products with alternative ingredients for fast growing broilers. *British Poultry* Abstracts. in press.

Houdijk, J. G. M., Smith, L. A., Tarsitano, D., Tolkamp, B. J., Topp, C. E. F., Masey O'Neill, H. V., White, G., Wiseman, J., Kightley, S. and Kyriazakis, I. 2013. Peas and faba beans as home grown alternatives for soya bean meal in grower and finisher pig diets. Pages 145–175 *in Recent Advances in Animal Nutrition.* Garnsworthy, P.C., Wiseman, J., Eds. Nottingham University Press, Nottingham, UK.

Ipema, A. F., Bokkers, E. A. M., Gerrits, W. J. J., Kemp, B. and Bolhuis, J. E. 2021a. Providing live black soldier fly larvae (Hermetia illucens) improves welfare while maintaining performance of piglets post-weaning. *Sci Rep* 11. https://doi.org/10.1038/s41598 -021-86765-3

Ipema, A. F., Gerrits, W. J. J., Bokkers, E. A. M., Kemp, B. and Bolhuis, J. E. 2021b. Live black soldier fly larvae (Hermetia illucens) provisioning is a promising environmental enrichment for pigs as indicated by feed- and enrichment-preference tests. *Appl Anim Behav Sci* 244.

Ipema, A. F., Gerrits, W. J. J., Bokkers, E. A. M., van Marwijk, M. A., Laurenssen, B. F. A., Kemp, B. and Bolhuis, J. E. 2022. Assessing the effectiveness of providing live black soldier fly larvae (Hermetia illucens) to ease the weaning transition of piglets. *Front Vet Sci* 9.

Janssen, R. H., Vincken, J.-P., van den Broek, L. A. M., Fogliano, V. and Lakemond, C. M. M. 2017. Nitrogen-to-protein conversion factors for three edible insects: Tenebrio molitor, Alphitobius diaperinus, and Hermetia illucens. *J Agric Food Chem* 65:2275–2278.

Jonas-Levi, A. and Martinez, J. J. I. 2017. The high level of protein content reported in insects for food and feed is overestimated. *J Food Compos Anal* 62:184–188. http://dx.doi.org/10.1016/j.jfca.2017.06.004

Kim, S. W., Less, J. F., Wang, L., Yan, T., Kiron, V., Kaushik, S. J. and Lei, X. G. 2019. Meeting global feed protein demand: challenge, opportunity, and strategy. *Annu Rev Anim Biosci* 7:221–243.

Komi, D. E. A., Sharma, L. and Dela Cruz, C. S. 2018. Chitin and its effects on inflammatory and immune responses. *Clin Rev Allergy Immunol* 54:213–223.

Liu, S., Wang, J., Li, L., Duan, Y., Zhang, X., Wang, T., Zang, J., Piao, X., Ma, Y. and Li, D. 2023. Endogenous chitinase might lead to differences in growth performance and intestinal health of piglets fed different levels of black soldier fly larva meal. *Anim Nutr* 14:411–424. Available at https://linkinghub.elsevier.com/retrieve/pii/S2405654523000665

Lu, A., Zhang, Q., Zhang, J., Yang, B., Wu, K., Xie, W., Luan, Y. X. and Ling, E. 2014. Insect prophenoloxidase: the view beyond immunity. *Front Physiol* 5:1–15.

Mellor, D. J. 2016. Updating animalwelfare thinking: moving beyond the "five freedoms" towards "A lifeworth living." *Animals* 6. https://doi.org/10.3390/ani6030021

Newton, G. L., Booram, C. V., Barker, R. W. and Hale, O. M. 1977. Dried Hermetia illucens larvae meal as a supplement for swine. *J Anim Sci* 44:395–400. Available at https://academic.oup.com/jas/article/44/3/395-400/4698064

OECD/FAO. 2023. *OECD-FAO Agricultural Outlook 2023–2032*. OECD Publishing, Paris, FR.

Oonincx, D. G. A. B. 2021. *Environmental impact of insect rearing.Pages 53–59 in Insects as animal feed: novel ingredients for use in pet, aquaculture and livestock diets*. CABI, UK.

Purschke, B., Scheibelberger, R., Axmann, S., Adler, A. and Jäger, H. 2017. Impact of substrate contamination with mycotoxins, heavy metals and pesticides on the growth performance and composition of black soldier fly larvae (Hermetia illucens) for use in the feed and food value chain. *Food Addit Contam Part A Chem Anal Control Expo Risk Assess* 34:1410–1420. https://doi.org/10.1080/19440049.2017.1299946

Rathore, A. S. and Gupta, R. D. 2015. Chitinases from bacteria to human: properties, applications, and future perspectives. *Enzyme Res* 2015. https://doi.org/10.1155/2015/791907

Šimůnek, J., Hodrová, B., Bartoňová, C. and Kopečný, J. 2001. Chitinolytic bacteria of the mammal digestive tract. *Folia Microbiol (Praha)* 46:76–78.

Smith, L. A., Houdijk, J. G. M., Homer, D. and Kyriazakis, I. 2013. Effects of dietary inclusion of pea and faba bean as a replacement for soybean meal on grower and finisher pig performance and carcass quality. *J Anim Sci* 91:3733–3741.

Spranghers, T., Michiels, J., Vrancx, J., Ovyn, A., Eeckhout, M., De Clercq, P. and De Smet, S. 2018. Gut antimicrobial effects and nutritional value of black soldier fly (Hermetia illucens L.) prepupae for weaned piglets. *Anim Feed Sci Technol* 235:33–42.

Spranghers, T., Ottoboni, M., Klootwijk, C., Ovyn, A., Deboosere, S., De Meulenaer, B., Michiels, J., Eeckhout, M., De Clercq, P. and De Smet, S. 2017. Nutritional composition of black soldier fly (Hermetia illucens) prepupae reared on different organic waste substrates. *J Sci Food Agric* 97:2594–2600.

Star, L., Arsiwalla, T., Molist, F., Leushuis, R., Dalim, M. and Paul, A. 2020. Gradual provision of live black soldier fly (Hermetia illucens) larvae to older laying hens: effect on production performance, egg quality, feather condition and behavior. *Animals* 10. https://doi.org/10.3390/ani10020216

van der Fels-Klerx, H. J., Camenzuli, L., Belluco, S., Meijer, N. and Ricci, A. 2018. Food safety issues related to uses of insects for feeds and foods. *Compr Rev Food Sci Food Saf* 17:1172–1183.

van Heugten, E., Martinez, G., McComb, A. and Koutsos, E. A. 2022. Improvements in performance of nursery pigs provided with supplemental oil derived from black soldier fly (Hermetia illucens) larvae. *Animals* 12. https://doi.org/10.3390/ani12233251

van Huis, A. 2013. Potential of insects as food and feed in assuring food security. *Annu Rev Entomol* 58:563–583.

van Huis, A. and Oonincx, D. G. A. B. 2017. The environmental sustainability of insects as food and feed: a review. *Agron Sustain Dev* 37. https://doi.org/10.1007/s13593-017 -0452-8

Veldkamp, T. and Vernooij, A. G. 2021. Use of insect products in pig diets. *J Insects Food Feed* 7:781–793.

White, G. A., Smith, L. A., Houdijk, J. G. M., Homer, D., Kyriazakis, I. and Wiseman, J. 2015. Replacement of soya bean meal with peas and faba beans in growing/finishing pig diets: Effect on performance, carcass composition and nutrient excretion. *Anim Feed Sci Technol* 209:202–210.

Widowski, T. M., Curtis, S. E. and Graves, C. N. 1989. The neutrophil:lymphocyte ratio in pigs fed cortisol. *Can J Anim Sci* 69:501–504.

Yu, M., Li, Z., Chen, W., Rong, T., Wang, G., Li, J. and Ma, X. 2019a. Use of Hermetia illucens larvae as a dietary protein source: effects on growth performance, carcass traits, and meat quality in finishing pigs. *Meat Sci* 158:107837. https://doi.org/10.1016/j .meatsci.2019.05.008

Yu, M., Li, Z., Chen, W., Rong, T., Wang, G., Li, J. and Ma, X. 2019b. Hermetia illucens larvae as a potential dietary protein source altered the microbiota and modulated mucosal immune status in the colon of finishing pigs. *J Anim Sci Biotechnol* 10. https://doi .org/10.1186/s40104-019-0358-1

Yu, M., Li, Z., Chen, W., Rong, T., Wang, G., Li, J. and Ma, X. 2020a. Evaluation of full-fat Hermetia illucens larvae meal as a fishmeal replacement for weanling piglets: Effects on the growth performance, apparent nutrient digestibility, blood parameters and gut morphology. *Anim Feed Sci Technol* 264. https://doi.org/10.1016/j.anifeedsci .2020.114431

Yu, M., Li, Z., Chen, W., Rong, T., Wang, G., Li, J. and Ma, X. 2020b. Hermetia illucens larvae as a fishmeal replacement alters intestinal specific bacterial populations and immune homeostasis in weanling piglets. *J Anim Sci* 98:1–13.

Zhu, M., Liu, M., Yuan, B., Jin, X., Zhang, X., Xie, G., Wang, Z., Lv, Y., Wang, W. and Huang, Y. 2022. Growth performance and meat quality of growing pigs fed with black soldier fly (Hermetia illucens) larvae as alternative protein source. *Processes* 10. https://doi .org/10.3390/pr10081498

www.ingramcontent.com/pod-product-compliance
Lightning Source LLC
Chambersburg PA
CBHW050538270326
41926CB00015B/3293